THE AIR FORCE IN SPACE

1959 - 1960

(U)

by
Max Rosenberg

June 1962
USAF Historical Division Liaison Office

FOREWORD

This is the second in a series of historical studies on the
role of the Air Force in space activities, prepared by the USAF
Historical Division Liaison Office. The first study, The Threshold of Space, 1945-1959, described the evolution of the national
space program with emphasis on the responsibilities of the Air
Force. For the present study, the author adopted a broad approach
that permitted timely coverage of USAF space policies, plans, and
programs within the context of both the national space program and
USAF strategic objectives. This office will also prepare compre-
hensive studies on particular portions of the space program.

The Air Force in Space, 1959-1960, is part of the History of
Headquarters USAF, Fiscal Year 1960.

This document is classified SECRET to conform with the classi-
fication assigned to sources of information used herein.

CHRONOLOGY OF USAF SPACE ACTIVITIES
FISCAL YEAR 1960

30 Jun 59 — President Eisenhower approved appointment of a National Aeronautics and Space Council (NASC) ad hoc committee to review and revise the preliminary national space policy issued 18 Aug 58.

5 Aug 59 — President approved the revised Basic National Security Policy calling for space exploration and development useful to U.S. scientific, military, and political endeavors, with emphasis on civilian (peaceful) activity.

13 Aug 59 — Discoverer V—launched successfully and all test objectives met except capsule recovery after ejection on 17th orbit.

19 Aug 59 — Discoverer VI—launched successfully and all test objectives met except capsule recovery after ejection on 17th orbit.

17 Sep 59 — Transit 1A navigation satellite—launched successfully but failed to orbit because of 3d-stage malfunction.

18 Sep 59 — Secretary of Defense Neil McElroy announced disapproval of the Army-Navy proposal for a joint military space operations agency "at this time." He also directed the transfer of management responsibility in the near future of the Samos reconnaissance and Midas early warning satellites from ARPA to the Air Force, of the Transit navigation satellite to the Navy, and of the Courier communication satellite to the Army. In addition, he directed the Air Force to assume total DOD responsibility for developing all military boosters, integrating payloads with boosters, and launching the combination.

21 Oct 59 — President directed the transfer of the "Von Braun team" and the Saturn superboosters from DOD (Army) to NASA. (He approved the transfer agreement on 2 Nov 59.)

7 Nov 59 — Discoverer VII—launched into orbit successfully but malfunctions prevented stabilization on orbit.

17 Nov 59 — OSD approved the transfer of management and development responsibility for Samos, Midas, and Discoverer from ARPA to the Air Force.

20 Nov 59 — Discoverer VIII—launched into orbit successfully but malfunctions prevented Agena engine shutdown at desired orbital velocity. Capsule ejected but not recovered.

15 Dec 59	Air Force Chief of Staff Gen Thomas D. White directed that relationships with NASA be kept at the highest possible level of harmony and cooperation.
12 Jan 60	NASC and the National Security Council (NSC) adopted a new national space policy, under revision since 30 Jun 59. Emphasis remained on civilian (peaceful) space activity but the military role received better recognition.
14 Jan 60	President formally requested Congress to amend the National Space Act of 1958 to clarify management responsibilities and streamline organizational arrangements. Under his proposal, the NASC and the Civilian-Military Liaison Committee (CMLC) would be abolished. (Although the House of Representatives enacted the President's suggested amendments in June, the Senate refused, preferring to await the recommendations of the new administration taking office in Jan 61.)
26 Jan 60	President approved the revised national space policy.
4 Feb 60	Discoverer IX—launched unsuccessfully because of a booster malfunction.
6 Feb 60	OSD approved transfer to the Air Force of management responsibility for ARPA's space-oriented applied research and component development projects.
19 Feb 60	Discoverer X—launching unsuccessful; range safety officer destroyed Thor-Agena combination at T+56 seconds.
26 Feb 60	Midas I—launched successfully but failed to go into orbit because of an Atlas-Agena separation malfunction.
13 Apr 60	Transit 1B—launched successfully into orbit. Also included first successful demonstration of an engine restart capability in space.
14 Apr 60	Chief of Staff White reaffirmed his directive that the Air Force seek the highest possible level of harmony and cooperation in its relationships with NASA.
15 Apr 60	Discoverer XI—launched successfully and all test objectives met except capsule recovery after ejection on 17th orbit.
2 May 60	Draft agreement made between DOD and NASA on establishment of the Aeronautics and Astronautics Coordinating Board (AACB), planned as the replacement of NASC and CMLC.
4 May 60	Navy reopened within JCS the question of a joint military space operations agency.

24 May 60   Midas II--launched successfully and about 75 percent of the test objectives met.

15 Jun 60   NSC approved its Operations Coordinating Board's revised Operation Plan for Outer Space. Based on the national space policy, the plan assigned some 35 "space" projects to the several participating Executive Department agencies.

16 Jun 60   Secretary of Defense Thomas S. Gates reaffirmed McElroy's 18 Sep 59 decision not to establish a joint military space operations agency. He also directed that unified and specificied commands exercise appropriate operational command over space systems.

22 Jun 60   Transit 2A--successfully launched into orbit. In addition a second satellite (for solar radiation measurements), carried "piggyback," was simultaneously placed into orbit. This was the first time that a single booster successfully orbited two satellites.

29 Jun 60   Discoverer XII--launched successfully but failed to orbit because of Agena malfunctions.

CONTENTS

THE AIR FORCE IN SPACE
1959 - 1960

Year III of the Space Age was marked by endless discussions and some
first steps aimed at restricting "space" to peaceful purposes. International agencies--governmental, scientific, industrial, and professional--
sought to expand the scientific exploration of space and widen the circle
of nations participating in mankind's newest and greatest adventure. By
the end of June 1960 the United Nations was at the point of establishing
a 24-nation Committee on Peaceful Uses of Outer Space with composition satisfactory to the two major space powers. The five Western nations at the
Geneva disarmament conference proposed a prohibition on launching and orbiting nuclear-armed satellites. At least seven countries expressed interest in joining the United States in cooperative space exploration programs,
and 10 nations in western Europe were in the midst of forming a joint space
agency for scientific purposes. Announcement of the impending establishment of the International Academy of Astronautics and of the Institute of
Space Law also furnished possible avenues for eventual peaceful agreement.
The American offer in December 1959 to let the Russians use the U.S. global
tracking system in support of Soviet man-in-space experiments was also a
hopeful sign, even though the Soviet Union did not accept the offer. And
already the American Telephone and Telegraph Company had firm plans to
employ a network of satellites for worldwide commercial telephone and tele-
vision purposes.[1]

Meanwhile, American and Russian attacks on the unknowns of space continued unabated. During 1959-60, American satellite and space probe experiments produced a steadily accumulating reservoir of knowledge and techniques. In the 12 months from 1 July 1959 through 30 June 1960, major American civilian and military space "shots" totaled 24, of which 14 were successful. The Russians announced three successful launchings in the same period. Also, numerous American probes of the lower-space environment, using Nike-Cajun, Aerobee, Javelin, Journeyman, and a host of other small rocket test vehicles, provided additional information.[2]

The American program was extremely successful in obtaining scientific knowledge and advancing space technology. The Russians, relying on the power of their larger-thrust boosters plus considerable scientific finesse, carried out far more spectacular feats. Hitting the moon and, later, photographing the far side of that natural satellite brought the Soviet Union great psychological and political prestige in the Cold War conflict with the Free World.

The importance of these feats and the benefits derived from them in the East-West controversies did not escape the Air Force. In fact, the Air Force role in the space adventures was perhaps the most important among American agencies. Yet, the Air Force had a far more vital and immediate concern with space. In a mid-December 1959 speech, Lt. Gen. Bernard A. Schriever, commander of the Air Research and Development Command (ARDC), warned that development of a military operational capability in space was much more than an adventure. Spectacular scientific space feats and the accompanying prestige had indeed become important in today's world, but they were not the kind of accomplishments upon which the survival or even the security of the nation could be founded. "My really

pressing concern," said Schriever, "is the direct and immediate importance
of exploiting the advantage that space offers to our vital military deterrent posture." This was an expression of the Air Force view that so long
as the possibility of war existed the United States had to exploit space
to the fullest in the defense of the nation.[3]

National Space Policy and Program

A basic weakness of the preliminary national space policy, promulgated
on 18 August 1958, had been the virtual omission or exaggerated deemphasis
of the potential military role in space. This undoubtedly reflected the
Administration's position taken at the opening of the Space Age and very
clearly stated by President Eisenhower on 2 April 1958: "A civilian set-
ting for the administration of space functions will emphasize the concern
of our nation that outer space be devoted to peaceful and scientific purposes."[4]

On 30 June 1959, President Eisenhower approved the appointment of a
National Aeronautics and Space Council (NASC) ad hoc committee to review
this preliminary policy. The group would recommend necessary revisions
in national space policy in the light of scientific, political, and military requirements, the National Space Act of 1958, the establishment of
the National Aeronautics and Space Administration (NASA), and other recent space developments. The revision, requiring more than six months to
formulate, was adopted at a 12 January 1960 joint meeting of NASC and the
National Security Council (NSC) and approved by the President on 26 January.[5]

Guidance for the formulation of the new space policy came primarily
from a more fundamental document--the Basic National Security Policy (BNSP)

approved by the President on 5 August 1959. The BNSP had called for an exploration and development program useful to our scientific, military, and political endeavors and, hopefully, to our efforts to regain recognition among the nations as the leader in space. It also stated that the military role was limited pretty much to use of the advancing technology to enhance military capabilities. There could be no invasion by the military of NASA's wide area of responsibility.[6]

Based on this guidance, the 26 January 1960 space policy acknowledged the tremendous significance and implications of a successful space program, especially in matters of international prestige involving U.S. competition with the Soviet Union. The Administration also acknowledged that the Russians led in the field and that restoration of position and prestige depended on U.S. ability to overtake them in terms of space payloads. The policy statement foresaw "great" possibilities for civilian application but noted only limited military activity within the next few years.[7]

Using the new space policy as its guidance, the Operations Coordinating Board (OCB) prepared a revised Operations Plan for Outer Space to replace the version in existence since 18 March 1959. The new plan, approved on 15 June 1960, restricted its outlook to the immediate future. It repeated the general policy objectives; listed various areas for research and exploration, for operational application, and for "international relations" consideration; and assigned these areas plus some 36 specific projects to one or more agencies--NASA, Department of Defense (DOD), Atomic Energy Commission (AEC), and the State Department. In keeping with the Administration's position, the plan outlined a conservative and orderly course of action.[8]

Although the revised space policy and operations plan constituted little change in the general outlook and direction of the Administration, a concurrent development had accorded increased recognition to the military role. On 14 January 1960 the President had proposed to alter the organization and management of the national space program and asked Congress to amend the National Space Act of 1958. The President wanted to obtain a clear understanding that a single national program, inherent under the act, was neither feasible nor desirable. Rather, two distinct programs had evolved. He also thought the transitional period, during which certain space projects had been transferred from DOD to NASA, had ended, and therefore he no longer needed to engage in detailed program planning, as called for by law. On this basis, he proposed to abolish NASC, since its function was to advise the President on space matters. Eisenhower also suggested elimination of the Civilian-Military Liaison Committee (CMLC),* deeming it inappropriate for Congress to prescribe the procedure under which DOD and NASA consulted and kept each other informed. Finally, to avoid duplication in the development of costly launching vehicles required by the two programs, the President asked for specific authority to assign development responsibility for each type of vehicle, regardless of its intended use, to either NASA or DOD.[9]

The Administration's space program and reorganization plans came under close congressional scrutiny, not only because Congress and the press

---

*The committee consisted of a chairman appointed by the President, one or more representatives from OSD, one or more representatives from the Department of Army, Navy, and Air Force, and an equal number of representatives from NASA.

believed that space progress had not proceeded rapidly enough but because the Administration did not consider the nation to be in competition with the Soviet Union or even admit that the United States lagged behind, except in booster capacity. The fact that 1960 was an election year undoubtedly added fuel to the smoldering controversy.

As early as July 1959 the Senate Committee on Aeronautics and Space Sciences criticized the lack of a well-defined space program and called for a "great deal more effort" to coordinate the military and civilian programs. On 29 October, Overton Brooks, chairman of the House Committee on Science and Astronautics, anticipated things to come when he announced that, upon the reconvening of Congress in January 1960, "it shall be our aim to probe every facet of the program to determine (1) why this nation is lagging behind in the exploration of space and (2) what steps can be taken to place the United States where it belongs, in the forefront."[10]

Hearings before several congressional committees started in January 1960 and lasted through the spring. Administration spokesmen, both civilian and military, generally supported the President's space policies and plans, although frequently conceding that, like it or not, the United States was in a race with Russia, it was lagging behind, and U.S. prestige was suffering. Just as frequently, the President denied the shortcomings and defended his space objectives.

A galaxy of outstanding government leaders and scientific experts debated long and arduously the question of one program, two programs, or, as some claimed, the lack of any program. The management and organization structure to be employed for one or for two programs was also examined. Since there appeared to be no clear-cut solution to the problems of one _versus_ two programs or of civilian _versus_ military control, the philosophy of two space programs enunciated by the President continued.[11]

The hearings also brought out specifics of the space programs. The House of Representatives accepted the Administration's request to abolish the NASC and CMLC, but before acquiescing it virtually forced the Administration to establish a replacement for the CMLC. Fearing the lack of a systematic means of coordination and information exchange between NASA and DOD, the House obtained from Deputy Secretary of Defense James H. Douglas what appeared to be an effective, if hastily concocted, substitute--the Aeronautics and Astronautics Coordinating Board (AACB). Composed of representatives from the two agencies and co-chaired by NASA's Deputy Administrator and DOD's Director of Defense Research and Engineering (DDR&E), the new unit possessed a major advantage over the CMLC. Since the AACB contained decision-making members from NASA and DOD, its decisions had the effect of directives upon the two parent organizations whereas CMLC could only "advise" and "recommend."[12]

Although the House approved the space act amendments in June 1960, the Senate took no action, preferring to await recommendations of the new administration, due to take office in January 1961. In practice, nevertheless, the NASC and CMLC ceased functioning after January 1960, and in May the AACB took on a provisional status. Some AACB panels were organized in June, but formal creation of the board did not occur until September 1960.*[13]

During the 1960 congressional hearings NASA released a plan for its exploration, research, development, and peaceful-application objectives

---

*Establishment of the AACB did not require new legislation.

covering the next decade. The plan outlined a series of 25 to 30 major shots per year of ever increasing complexity, with a prediction of a steadily mounting expenditure of funds during those years.[14]

The Defense Department, on the other hand, disclosed no such plan. The Advanced Research Projects Agency (ARPA), sole director and manager of DOD's space program from its establishment in February 1958 until November 1959, had drawn up in July 1959 a long-range research and development plan. After obtaining comments from JCS and the services, DDR&E, the final authority on DOD development matters, neither approved nor implemented the plan and finally declared it obsolete.

DDR&E explained to Congress that it had taken this action on the assumption that DOD was not interested in space flight and exploration as ends in themselves but rather in the application of flight in space as a means to a more general end—the defense of the United States and its allies. Therefore, DOD space efforts would be considered only as an integral part of the overall defense program to enhance military capabilities. This effort would be restricted to one of two objectives: development of systems in which the use of space flight would enhance the defense, or the development of components for advanced systems that would probably become essential as the application of space technology to defense became better understood. On this basis, DDR&E claimed that it was not logical to formulate a long-range military space program separate and distinct from the overall defense program. As a result, the Air Force showed no space program when presenting its budgetary requests to Congress but listed the Samos reconnaissance system under strategic developments, the Midas early warning system under air defense developments, and so on.[15]

Since there were two space programs and severe limitations on DOD's area of responsibility, Congress was especially interested in preventing undesirable duplication and insuring free and proper development and interchange of information, materiel, and facilities between DOD and NASA. This posed an extremely serious problem in the light of the President's request to abolish the two advisory and coordination bodies (NASC and CMLC) and to meet certain peaceful and military requirements by incorporating them in single space projects, such as a passive communication satellite, a meteorological satellite, a geodetic satellite, and of course, certain advanced boosters.

In a steady stream, top Administration and military officials testified that all was well. The use of many committees, panels, and old-fashioned telephone and on-the-spot discussions made for an excellent working relationship between NASA and DOD. Requirements of both agencies were carefully evaluated and schedules jointly prepared. Range facilities were equitably shared on a priority basis. DOD transferred specialized military personnel to NASA on request, even when their retention within a particular military service was highly desirable. Both NASA and DOD pointed to the delineation of missions and claimed that, by and large, they had arrived at a satisfactory demarcation. The few gray areas in question were settled by conference. If this failed, the matter passed to the President for decision. The testimony was impressive, but when Congress persisted in demanding some formal body and set of rules to insure the continuance of this good state of affairs, the Administration quickly acceded and set up the aforementioned Aeronautics and Astronautics Coordinating Board.[16]

At first glance, the President's budget request of January 1960 for the space program seemed to contain significant increases over that of fiscal year 1960. Closer examination disclosed, however, that in the case of NASA the increase was really quite modest in the light of the recent transfer of the large Saturn superbooster program to that agency.[*] A subsequent supplemental request of more than $100 million specifically for Saturn and Nova, also a superbooster development, quieted much of the congressional criticism. The DOD budget request for space contained only a slight increase. In this instance, Congress, strongly motivated by the unfortunate U-2 events of May 1960 and the consequent loss of this means of intelligence-gathering, appropriated amounts sizably greater than those requested for the Midas and Samos projects.[17]

In summary, the Administration had ably defended its policies, plans, and programs for space and in large degree retained them in much the fashion they were presented to Congress.

<u>Air Force Space Policy</u>

As far back as 1946, the Air Force had expressed an interest in space. By January 1948, it had "staked out a claim" in that environment by promulgation of a policy statement and through the intervening years continued to maintain the position. However, by 1959 the policy was of questionable worth, other than for internal consumption, and the extensive USAF effort devoted to space was almost entirely under the financial sponsorship, management, and technical direction of ARPA or NASA. This state of affairs the Air Force had persistently but unsuccessfully attempted to

---

[*]See below, pp 15-16.

alter, but so long as it existed, a statement of policy had little meaning and USAF hopes and aims remained beyond attainment.

During September-December 1959, this awkward situation was somewhat alleviated when Secretary of Defense Neil McElroy assigned to the Air Force responsibility for the development and operation of all DOD boosters and several space systems. In February 1960 the Air Force also received authority to take over a major segment of the ARPA-sponsored space study and component development program. These actions provided impetus to formulate an official Air Force space policy.

General Doctrines

The basic tenet of Air Force policy declared that space was simply a location—not a function or a military program. Equally important, no real dividing line separated "air" and "space," and the total expanse beyond Earth's surface constituted one vast operating arena—aerospace. Throughout its history, the Air Force had constantly pushed for greater speeds and higher altitudes in its weapon systems because these characteristics increased military effectiveness. Military expansion into space was therefore not so much a challenging adventure as a vital and essential step in insuring the nation's future security.

A second major tenet was that the USAF mission in the vastness of aerospace could be fulfilled without regard to whether a weapon system was aeronautical, astronautical, or a combination of both. The prime criterion in the selection of a system to satisfy a military requirement should be its effectiveness. Thus the Air Force would consider development and operation of a space system only if it were the sole means of doing a required job, if it were the best way to do that job and not prohibitively expensive, or if it were the most economical way to do the job. On this basis, the Air Force foresaw for many years to come an operational force consisting of weapon systems—aircraft, missile, and space—selected to meet essential military requirements. Space for its own sake was not a suitable consideration.

A third basic tenet of Air Force policy concerned the objectives of the national space effort. The Air Force supported the President's "Space for Peace" program for the benefit of all mankind, seeing in this no particular contradiction to its policy of safeguarding the peace by maintaining a strong military capability in space. "Aerospace Power is Peace Power," the Air Force considered a truism, and until such time as peace could be guaranteed by other means, there were legitimate requirements for a military space program.[18]

USAF policy did not officially contain these doctrines other than by indirection. Air Force Manual 1-2, United States Air Force Basic Doctrine, published on 1 December 1959, outlined objectives by definition and implication without actually employing the term "space." On the other hand, there was nothing secretive about the policy, and USAF civilian and military officials used its substance continually in testimony before congressional committees during the early months of 1960.[19]

Air Staff and field command elements had wanted a formal statement of policy for some time, but top Air Force leaders wisely tended to tread softly because of the current DOD space management and organization structure. In July 1959 the Deputy Chief of Staff for Plans and Programs (DCS/P&P) began formal study of a space policy. This work quickened after the transfer from ARPA of management responsibility for certain space projects, in November. By the end of December a number of coordinated Air Staff statements on pertinent portions of the space program were ready for the Chief of Staff's Policy Book and were used in the congressional hearings.[20]

In December 1959, General White asked for an all-inclusive statement suitable for official dissemination. DCS/P&P filled the request on 17 January 1960, but subsequent review elsewhere in the Air Staff and in the Office of the Secretary of the Air Force (OSAF) produced some revisions. A final version containing substantially the basic doctrines discussed above went to the Chief of Staff on 14 March, but it was shelved to await a more propitious time for issuance. USAF leaders feared that publication of an official policy statement at a time when so many facets of the space program were still undecided would have unfavorable reverberations in Congress, the Office of the Secretary of Defense (OSD), and the other military services.[21]

Policy on NASA

Top Air Force leadership held that the National Space Act of 1958 provided an adequate framework for the effective advancement of space technology. No conflict existed between the aims of NASA (space exploration and civilian applications) and those of DOD (military applications), and the Air Force intended to promote a high degree of cooperation with NASA. However, in the interest of efficiency, the Air Force wanted a single point of contact between NASA and DOD and considered itself as the logical DOD agency.[22]

Air Force support of this policy went far beyond "lip service." In a letter to the Air Staff on 15 December 1959, General White stated, "I would like every member of the Air Force to do everything within his power to maintain the same degree of harmony and cooperation with NASA ⟨as had existed with the National Advisory Committee for Aeronautics⟩." He repeated this admonition even more strongly in a 14 April 1960 message to his deputies for personnel and development: "I want to make it crystal clear that the policy has not changed and that to the very limit of our ability, and even beyond it to the extent of some risk to our own programs, the Air Force will cooperate and will supply all reasonable key personnel requests made on it by NASA."[23]

Air Force support of NASA in the personnel area became sufficiently noticeable to draw comment in newspaper articles. As early as 13 December 1959, the Washington Evening Star headlined an article "Space Agency Fills Top Jobs with Brass from Air Force." By the end of the fiscal year the Air Force had assigned 69 military personnel to NASA. Although many were AFROTC lieutenants, others wore stars. In addition, USAF military and civilian personnel served on many NASA advisory committees, an arrangement designed to be mutually advantageous.[24]

The Air Force also furnished extensive technical support to NASA's Mercury man-in-space project, including delivery and launching of Atlas boosters, use of Air Force range instrumentation and tracking facilities, utilization of aeromedical equipment and experimental findings, and, of course, assignment of three of the seven astronauts-in-training. The Air Force also participated actively on a contractor basis in virtually every NASA satellite and deep-space probe.[25]

This policy avowedly served the Air Force as well as NASA. General White in April 1960 conceded this when he noted:[26]

I am convinced that one of the major long range elements of
the Air Force future lies in space. It is also obvious that NASA
will play a large part in the national effort in this direction
and, moreover, inevitably will be closely associated, if not eventually combined with the military. It is perfectly clear to me
that particularly in these formative years the Air Force must, for
its own good as well as for the national interest, cooperate to
the maximum extent with NASA . . . .

Policy on Development

As discussed above, the Air Force held the view that space and space
systems should not be considered separate from the entire range of weapon
systems. "Specialized concepts, organizations and procedures should not
be developed for the pursuit of so-called 'space' programs" was a major
USAF principle.[27] This long-held stand produced its first results on 18
September 1959, when McElroy authorized the transfer of certain projects
from ARPA to the military departments. The Air Force obtained Samos and
Midas, as well as full responsibility for the development and launching
of all DOD boosters. In November, the actual shift took place and also
included the Discoverer research satellite. This transfer was the first
step in removing ARPA from control over DOD's space program. On 30 December 1959, Dr. Herbert F. York, DDR&E, restricted ARPA activity to specific fields of "advanced research" in ballistic missile defense, solid-
rocket propulsion, materials, and the like. Early in February 1960,
York approved the shift to the Air Force of much of ARPA's space study
and component development program. Several days later he informed a
House committee that "it /ARPA/ no longer does play a role in the space
program."[28]

The assignment of all "defense" boosters to the Air Force brought up
the question of disposition of the ARPA-sponsored, Army-developed Saturn

superbooster. McElroy's directive implied that the Air Force would assume both the ARPA and Army roles, and the Air Staff immediately asked for reassignment of the Army Ballistic Missile Agency (ABMA), which was developing Saturn. However, because of lack of funds and specific military requirements, there was considerable opposition in OSD to continued development of Saturn under DOD auspices. Since the Administration also was concerned with the financial problems, the President directed that the question of a single superbooster development agency be settled once and for all.[29]

On 8 October 1959, Thomas S. Gates, Jr., Deputy Secretary of Defense, asked JCS to comment on the selection of NASA or DOD as the development agency for Saturn and to appraise the validity of military superbooster requirements. JCS favored DOD as the responsible agency, claiming it was significantly better equipped for the job, but JCS could not cite any immediate requirement for a superbooster, noting that this need lay a few years in the future.[30]

The matter was settled on 21 October at a meeting of the President and his top advisers. The President directed that ABMA's Development Operations Division (the Von Braun team) and Saturn be transferred to NASA, and on the same day, DOD and NASA drafted a joint agreement explaining the Administration position. When Gates asked JCS to comment on the agreement, JCS reaffirmed an eventual need for superboosters and asked for a revision of the agreement, which implied stronger civilian and military space programs would result. "Actually," said JCS, "this transfer strengthens the civilian agency but at the expense of the military effort by removing both ⌐ sic ⌐ a facility, a program, and key personnel from military direction." Gates refused the requested revision.[31]

When the President initialed the 21 October agreement on 2 November 1959, he, in effect, shut the door for the second time on USAF aspirations to develop military superboosters to meet obvious requirements of the future. The restriction placed the military services somewhat at the mercy of a civilian agency whose objectives and outlook were quite different from their own. The Air Force feared that this major policy decision on space development also endangered the future of the Dyna-Soar manned spacecraft.[32]

Air Force space development policy called for concentration of responsibility in one service. Quite naturally the Air Force deemed itself the logical service since it possessed, after September 1959, the entire field of DOD boosters as well as the major share of the experience and facilities employed in the military space program. This in no sense meant that the Air Force wanted to eliminate the other services from space. The Air Force realized that the Army and Navy had valid requirements for the benefits derived from space systems and that they possessed specialized capabilities and facilities valuable in the conduct of the space effort. In the final analysis, the Air Force wanted a single point of development and management responsibility where requirements of all services could be evaluated and met and where the capabilities of all could be utilized most efficiently and economically.[33]

Within its headquarters, the Air Force practiced the policy it preached. It established no special space organization. Rather, all elements of the space program were kept in functional channels and the respective deputy chiefs of staff played their usual roles in policy-making, planning, development, production, operations, and the like.[34] This was a vast improvement over the indecision and vacillation that had characterized management of the guided missile program. On the other hand, the experience and knowledge obtained in the missile field paved the way for continuance of more normal procedures in the newly opened space field.

Policy on Operations

In keeping with the principle that space systems were only improved means of accomplishing certain aerospace missions, the Air Force contended that space operations should be conducted under the same procedures applied to other weapon systems. The using service would support a space system administratively and logistically, and one of its functional commands would operate it under the control of a JCS unified or specified command. This would avoid the creation of additional DOD agencies. Information obtained from operating space systems would be disseminated to all interested agencies through existing channels.[35]

This concept of operations twice came under attack during fiscal year 1960. Late in April 1959, Adm. Arleigh A. Burke, Chief of Naval Operations, pointing to the "very indivisibility of space," the projected large-scale aeronautical operation, and the interests of all three services in space, proposed that JCS create an agency under its aegis to coordinate all space "facilities and functions." Gen. Maxwell D. Taylor, Army Chief of Staff, quickly concurred, but General White was opposed, claiming that space systems only represented a more effective means of accomplishing a mission and should be assigned to the appropriate unified or specified command.[36]

Late in May, McElroy entered the controversy by asking JCS for its recommendations on assigning operational responsibility for four systems

being considered for transfer from ARPA to the military departments. He
also wanted to know what other agencies had an interest in the data to be
obtained by the four systems. Finally, McElroy wanted suggestions on
which service should support the systems logistically and undertake system
improvements.[37]

The Joint Staff and JCS spent the next two months attempting to reach
an agreement on the Burke proposal and on recommendations to McElroy.
Until 17 July 1959, when JCS agreed to submit divergent views, the Army
and Navy, joined by the Joint Staff, adamantly called for creation of a
joint military space operations command. On the question of logistics and
product improvement, they recommended an arbitrary division of the systems
without regard to roles and missions or past development activities. The
Air Force steadfastly maintained its original stand, and split views went
to McElroy on 24 July.[38]

McElroy met with JCS on 13 August 1959 and made it known that his
thinking on the matter followed rather closely the Air Force policy. Nevertheless, when asked to comment on a 28 August draft of McElroy's decision,
the JCS members again failed to reach an accord. On 3 September, McElroy
met with JCS, the three service Secretaries, and Dr. York. He reiterated
his tentative decision of 28 August and stated his willingness to accept
only revisions that clarified ambiguities in language; the substance was
inviolate.[39]

McElroy formally released his decision on 18 September 1959. The part
dealing with the operational question stated that establishment of a joint
military organization "does not appear desirable at this time." He pre-
ferred the current organization in order to realize full advantages from

existing support capabilities. In the interest of economy and efficiency, he delegated operation of all DOD boosters to the Air Force, along with the task of all system integration—in simplest terms, the mating of booster and payload.

Other provisions called for the early transfer of management responsibility for Samos and Midas to the Air Force, for Transit to the Navy, and for Courier to the Army. Pending actual transfer, the respective departments were to prepare operational plans for each system, including contemplated user relationships with unified and specified commands. McElroy agreed to obtain JCS comments on the plans before approving them.[40]

The Navy, dissatisfied with McElroy's decision, reopened the question on 4 May 1960, shortly after the Air Force submitted its operational plans for Samos and Midas. Burke listed several reasons for again asking for a joint organization to control space systems and the military forces and facilities supporting them. These included the rapid technological advances of the last half-year that had brought several systems to the "operational threshold," the establishment of a large interservice support group for recovery of Mercury capsules, and the pending creation of joint agencies for command and control and for communications.[41]

Although JCS agreed on 18 May 1960 to send divergent views to Gates, Secretary of Defense since December 1959, these were not ready until 31 May. The Army and Navy supported the reasons listed by Burke; the Air Force simply held that basically the situation had not changed since issuance of the September 1959 directive. Gen. Nathan F. Twining, Chairman, JCS, delayed dispatch of the conflicting views and informally discussed the question with the Secretary of Defense.[42]

On 16 June 1960, Gates informed JCS and the three military departments that he had reaffirmed the 18 September 1959 decision. "Additionally," he went on, "it is desired to emphasize that the establishment of a joint military organization for control over operational space systems does not appear necessary or desirable at this time." He altered the earlier directive only in one major aspect, specifically directing that unified or specified commanders would exercise appropriate control when a space system became operational.[43] Thus, in the course of less than a year, Air Force policy on operational responsibility twice underwent scrutiny and was adjudged suitable.

Summary

In retrospect, it appears that Air Force policy on space, although always under constant revision, was effective. Cautiously drawn and prudently publicized, the policy provided logical guidelines for USAF efforts to dominate the "space picture." But by July 1960 the Air Force was still a long way from that goal. Several segments of the DOD space program remained unassigned, and uncertainty as to their assignment still existed. Nevertheless, in those areas in which the Administration and OSD had reached a decision, the Air Force had carried away the lion's share. This was a great improvement from a year earlier, when the Air Force had virtually no space activity that it could call its own.

Air Force Space Program

The thesis of the indivisibility of air and space imposed on USAF aerospace planners a sort of Jekyll-and-Hyde complex. In order to outline concepts of operation, military capabilities, and development programs, the planners had to plan in terms of space as a separate medium. This, in turn, led to conflicts with basic policy, circuitous semantics, and indecision in planning. The unknowns of space also added to the complexity of the problem. As a result, few official pronouncements to guide and educate Air Force commanders and personnel at large were forthcoming during 1957-60.

Planning for Space

As early as 1958 the Chief of Staff had commented on the desirability of issuing preliminary long-range concepts for space operations through the medium of an Air Force Objective Series (AFOS) paper. Initial drafts proved too broad, too uncertain, and, in a sense, too much at odds with current policy.[44] A third draft, prepared in September 1959 after letting the matter rest for over a year, covered the 1960-70 period for both peacetime and wartime operations in space. It outlined the expected USAF roles and stated broad weapon system objectives, with priorities for strategic offensive and defensive purposes, support functions, and reconnaissance duties. It also proposed to emphasize the development and use of manned systems as the most efficient way to carry out space operations.[45]

The 1959 draft of the objectives statement received a cool reception during informal Air Staff coordination. "Policy" officials within DCS/ Plans and Programs decried treatment of space as a separate entity. Their "plans" counterparts suggested that publication be delayed pending inclusion of more meaningful technical guidance expected to be available shortly from ARDC.[46]

Since subsequent drafts were equally unacceptable, the Long Range Objectives Group of the Directorate of Plans sought a new approach. On 19

February 1960 the group admitted to the Directorate of Development Planning that "this office has been unsuccessful in getting an AFOS paper on space off the ground" and asked the development planners to take "a cut" at the problem. The statement had to be conceptual in nature, providing a firm point of departure for a statement of space operational requirements and desired space systems.[47] By the end of June 1960, the Directorate of Development Planning still had the matter under study.

Theoretically, the AFOS document would supply broad objectives on space operations. The Air Staff expected another document, the required operational capability (ROC), to provide the "how" for meeting these objectives. Unfortunately, the ROC under preparation by the Directorate of Operational Requirements faced conditions not unlike those of the AFOS, and it also failed to gain approval during the year. The ROC did not completely complement the AFOS drafts, both being studied concurrently, and the former had little in the way of formal guidance from the latter.

By April 1960 the Directorate of Operational Requirements had completed its first draft of a 23-page statement detailing the military capability deemed essential for aerospace systems one to two decades in the future. In its simplest terms, the ROC called for an operational space force per se. The "breaching of the gravity barrier" would remove current limitations imposed by the physical environment on deterrent forces. Only the lack of vision and audacity would prevent attainment of an ideal deterrent --a self-sufficient, manned military space force with multimission capabilities, dispersed in the cislunar and translunar regions. Elimination of this force by the enemy would be prerequisite to attacking the nation, and the outcome of the aerospace battle could well be decisive without involving surface forces.

The ROC deplored the fact that the current program of research and de-velopment was still tied to operation from the surface of Earth. "Revolu-tional" rather than evolutional development was deemed mandatory. Contin-ued reliance on evolution, the Directorate of Operation Requirements con-tended, could achieve only second place in military technology and, conse-quently, in military power. The United States needed selective "quantum jumps" or "leapfrogging" in virtually every area of development--propulsion, human factors, materials, and guidance. Such action would require major innovations at all policy levels, and the operational products had to be available by 1975.[48]

Several Air Staff agencies questioned these utopian requirements on several important points. Quantum jumps were expensive and risky to under-take. As the Assistant for Advanced Technology remarked, technological advance, particularly of a revolutionary nature, could not be ordered. Another criticism involved Air Force policy (and the national policy) that designated space as a location or medium. The ROC appeared to call for space weapons for the simple reason that it was now possible to operate in space and disregarded the fact that an Earth-based or aeronautical weapon might do the required job better or less expensively. But the Air Force had already decided that past standards should hold: development of weap-on systems would depend on military and cost effectiveness, not on aero-nautical or astronautical characteristics. Finally, the very important question of the necessity for manned space systems remained open to debate and resolution.[49]

In the summer of 1960, it was obvious that there would have to be a reconciliation of the many conflicting statements on policy, objectives, capabilities, and requirements before the proposed ROC for aerospace could gain official approval.

A third document, prepared by the Directorate of Development Planning, concentrated specifically on space development planning. Under normal circumstances, it would have followed in sequence and detail the objectives and capabilities called for in the AFOS and ROC documents. Since neither of these existed in approved form, the 120-page development planning note of October 1959 (revised in March 1960) outlined the qualitative force structure required in 1960-80, based purely on technological factors, and the research and development program required to provide that force. The planning note examined four critical areas: Air Force missions and Soviet technological capabilities, which in large measure determined operational requirements; the costs involved and the scope of action afforded by "reasonable" military budgets; possible limitations on the use of space during the period; and the requirements for man in space.

The projected force was phased in five-year spans, taking into account probable Soviet technological advances. The first important consideration was the defense of existing American strategic deterrent forces, as well as American population, property, and resources. As important, the Air Force had to develop an offensive force capable of surviving an attack and of retaliating in sufficient strength to destroy the enemy's will to fight. The Air Force must also develop space weapons capable of improving greatly the conduct of various reconnaissance and surveillance functions—early warning, strategic reconnaissance, strategic warning, mapping, and tactical reconnaissance. The last consideration covered support forces to provide control of communications, navigation, weather observation and operations.

Within each of the four critical areas examined, the planners laid out
the categories of a complete program of research and development. Some
portion of each category was already under development, but much remained
to be done. The categories of development included the following:

Operational development systems—declared technically and economically
feasible now

Operational development subsystems—declared technically and economically feasible and required for follow-on systems

Advanced systems and subsystems—currently under study to determine
technical and economical feasibility

Applied research—effort to seek new inventions

Exploration—preliminary investigation of new realms of flight

In the opinion of the planners, each category of effort assumed equal importance in the long-range program, and only constant attention to all
phases could assure ready attainment of the required operational force.

In conclusion, the planners noted that an immediate shift in emphasis
was required to obtain the goals of the proposed program in an orderly and
evolutionary manner. They did not imply, however, that this should constitute the whole development program. They intended only to set forth those
space development requirements essential to an adequate overall military
posture.[50]

Management and Organization of the Space Program

At the beginning of fiscal year 1960 the Air Force had no space program
that it could call its own other than the near-space Dyna-Soar project.*

*See below, pp 46-49.

To be sure, USAF research and development covered the whole range of space subjects from exploratory research to system development, but it was being done under the sponsorship, management, and direction of ARPA. In addition, the Air Force expended much of its time, resources, and energy supporting NASA activities.

The ARPA arrangement was far from satisfactory to the Air Force because ARPA enjoyed almost complete freedom in deciding which military requirements, stated by the services, would be pushed, combined, or ignored. Indirectly, it could set priorities on the portions of the space effort by determining the level of funding accorded each. Through its development assignments, ARPA could profoundly affect or color future operational roles. It possessed authority to go over, under, and around OSAF and the Air Staff in committing personnel, facilities, and other resources of the field commands, particularly ARDC and its units. Thus the Air Force retained responsibility for determining requirements to satisfy its assigned mission but had to persuade higher echelons to approve, fund, and assign the projects necessary to meet these requirements.

This unsatisfactory condition had been the subject of discussion on many occasions, and it arose again during August–September 1959 between Roy W. Johnson, Director of ARPA, and Joseph V. Charyk, Assistant Secretary of the Air Force (Research and Development). Charyk insisted that for efficient prosecution of the USAF research and development program all directives to Air Force units should be directed through and reviewed by him. Johnson thought that his management and technical responsibilities dictated direct contact with the field units.[51] The projected removal of ARPA from the space field, announced in September 1959, served in large part to settle this issue.

Although the transfer of space projects to the services tended to eliminate some of the out-of-channel procedures, it in no sense alleviated the
tight control exercised by OSD agencies. Lt. Gen. Roscoe C. Wilson, Deputy Chief of Staff for Development (DCS/D), feared that this continuation
and growth of civilian technical control had dangerous overtones. Not
only did it create an imbalance between technical and military influence
but it cost the Air Force many hours of briefings and much loss of time
and direction awaiting decisions from "on-high." Wilson warned that "this
trend toward project direction from Olympus" had to be solved. "Unless it
is reversed," he declared, "DOD and all services will bog down in red
tape."[52]

How to balance influence and decision-making between civilian and military officials was also a problem within the OSAF-Air Staff complex. In
October 1959, midway between the announcement of the pending transfer of
space projects and their actual shift, Secretary of Air Force James H.
Douglas directed that all space actions be taken "within the framework of
the AFBMC (Air Force Ballistic Missile Committee)." *[53] This concentrated
all decision-making prerogatives in civilian hands and markedly reduced
Air Staff participation.

Late in November, Charyk learned that the Air Staff intended to receive briefings from the Air Force Ballistic Missile Division (AFBMD) on
several space system development plans prior to their presentation to
AFBMC. He noted his concern and stated that Douglas had termed such a

---

*AFBMC had been established in November 1955 as the single decision-
making body for the ballistic missile program within the OSAF-Air Staff
complex. It consisted of the Secretary, Under Secretary, and Assistant
Secretaries of the Air Force plus one representative from the Air Staff.

briefing to the Air Staff without previous AFBMC review and comment as a "waste of time." But Generals Wilson and LeMay contended that it was necessary to arrive "at a corporate Air Staff position on Space Systems." They based their view on experiences in the ballistic missile program where similar restrictions had prevented coordinated Air Staff positions. In this instance, AFBMD representatives made an informal presentation to AFBMC and received the necessary guidance. When the revised plans were ready, Dudley C. Sharp, Secretary of the Air Force since 11 December 1959, allowed prior review by the Air Staff.[54]

In the spring of 1960, Sharp formally authorized a 90-day trial of a modified management structure within Headquarters USAF that noticeably increased Air Staff participation in space development plans. Although the Air Staff position could now be made known in advance, the decision responsibility still remained with AFBMC.[55]

When the Air Force regained management of a portion of the space program on 17 November 1959, Samos, Midas, and, unexpectedly, Discoverer, were transferred to it. In February 1960, York approved the shift to the Air Force of a major segment of ARPA's space study and component development program. During 1960 the Air Force made repeated attempts to obtain jurisdiction over the Tackle-Steer-Decree communication satellite projects, the Spacetrack-Spasur interim satellite detection and follow-on systems, and the antisatellite and ballistic missile defense studies and systems (Saint, Spad, RBS, and others). Since the Air Force was responsible for all DOD booster development and launchings, it played important roles in the Army's and Navy's space programs, particularly the Courier and Transit satellites.

In organizing its growing space program, the Air Force realized that the effort involved major risks and uncertainties as well as potentially large rewards. Only careful judgment could balance the priority of requirements, technical problems, and costs. As Charyk reported to Congress in February 1960, a maximum risk program was prohibitively expensive, wasteful, and inefficient, while a minimum risk program would cause unacceptable delay in attaining operational capabilities. Somewhere in between lay the proper balance in the light of military threats and requirements.[56]

Requirements would probably develop eventually for operations in the cislunar and translunar regions, but in the immediate future the Air Force had to concentrate on systems operating within several hundred miles of Earth's surface. Brig. Gen. Homer A. Boushey, Director of Advanced Technology, aptly summed up the Air Force space outlook with a mixed but pertinent metaphor: "We can best go into space with our feet firmly planted on the ground." The Air Force would encourage the widest use of imagination in propounding a program but employ conservatism and "horse-sense" in developing it.[57]

In the light of its experience, the Air Force divided the space development effort in fiscal year 1960 into three major areas. The first—pure studies—sought new ways of doing military jobs and outlined possible system approaches. The second included applied research for the purpose of developing techniques that provided essential ingredients for future systems. Most important here was propulsion—"the key to space use" was General Wilson's descriptive remark. Larger thrust and lower cost were essential to reduce the tremendous pound-per-dollar expense of boosting spatial payloads and insure continuous large-scale space operations. The

third area—system development—was the final goal, the last step in the
study-research-development process to meet requirements stated years earlier. Equipped with the products from this cycle, the Air Force could
effectively conduct its offensive, defensive, reconnaissance, surveillance,
and support operations.[58]

Samos-Midas-Discoverer

Throughout the year, planning and programming for the Samos reconnaissance satellite, the Midas early warning satellite, and the Discoverer research satellite were lumped together. Historically, the three had grown
out of a proposal partially outlined as early as 1946 and established as
a system development nine years later. Samos and Midas depended greatly
on Discoverer accomplishments and both expected to use the same orbital
vehicles and many common ground support facilities.

Conflicting decisions and indecision marked the Samos-Midas-Discoverer
program during fiscal year 1960. Virtually every phase—technical, operational, funding, requirements—remained in a constant state of flux.
Disagreement and disputes between civilian and military experts in the
technical area, between OSD and the Air Force in funding, and between the
Air Force and other services in the operational area kept the program in
continuous turmoil. At year's end, certain hopeful signs indicated that
some order might soon evolve.

In a portent of things to come, ARPA informed the Air Force on 23 June
1959, that demands from other space projects would substantially reduce
the funds previously planned for Samos and Midas. At the same time, ARPA
directed a major technical reorientation of Samos and failed to approve
the second phase of Midas development.[59] Thereafter, the Air Staff and

the field commands were kept busy preparing, revising, and defending a series of development and operational plans for these satellite systems.

The reliability, life, complexity, and priority of several Samos and Midas subsystems were in question. Some OSAF and OSD officials and Administration scientific advisers doubted Air Force ability to obtain acceptable success in many of the areas. Moreover, they envisioned substantial savings by using less sophisticated equipment and subsystems requiring less costly and complex ground facilities. Accordingly, a "fly before you buy" view took hold.[60]

Military leaders, faced with critical operational requirements and with pressures from potential users, maintained a more optimistic position. They felt that reliability, long-life, simplicity, and other desirable performance characteristics would be obtained as development and testing proceeded. They naturally indorsed the concept of concurrency previously employed in the ballistic missile program as a means of obtaining early operational capabilities for Samos and Midas.[61]

The differences in outlook caused an almost continuous review of the technical approach and funding. During the course of the year, AFBMD prepared at least five development plans for Samos, four for Midas, and several for Discoverer. None were completely satisfactory to civilian authorities in OSAF and OSD, either for technical or financial reasons.[62] A development plan theoretically outlined a year's course of action and projected it over the next few years. Lacking an approved plan, USAF development officials and contractors could carry on only in uncertain fashion.

OSD and OSAF technical experts claimed that military leaders were so preoccupied with operational considerations that they were shaping development requirements without realistically appraising the current state of the art. They also felt that military leaders wanted to proceed prematurely with development and construction of complex operational bases and stations.

Repeatedly refused the use of operational funds, the Air Force proposed what it called a "development/operational plan" to cover the transition period from development to operational status. Initially, these plans received a cool reception from OSD officials, but with the downing of the U-2 over the Soviet Union on 1 May 1960 and the subsequent international political repercussions, the Air Force sensed the likelihood of increased support for the Samos-Midas-Discoverer program.[63]

Intensive activity within OSD and OSAF during May and June culminated in an Air Staff directive for preparation of a plan to exploit as early as possible any intelligence data that might be obtained from Samos flight tests. It soon became apparent that the project would shortly receive special Administration attention and be reestablished under special management.[64]

During the spring of 1960, Congress became intimately involved in Samos-Midas progress. In light of the U-2 incident, Congress called for the rapid development of both space systems and voted sums far in excess of the Administration's requests for fiscal year 1961.[65]

Uncertainties and indecision on the technical and budgetary aspects of Samos and Midas affected planned operational dates for the two space systems. In July 1959 the Air Force was aiming for an initial operational capability date of October 1961 for both systems. For Midas, the scheduled date subsequently slipped to July 1962, to January 1963, to April 1963, but

York predicted before a congressional committee in March 1960 that the satellite would not be operational before 1965. Samos dates also slipped, to July 1962 and then July 1963.[66]

Discoverer displayed a good capability to carry out all but one of its tasks. A total of eight launchings occurred during the year, and five Discoverers went into orbit. Since none of the orbiting capsules were recovered, flight tests were suspended during several extended periods while the recovery subsystem underwent minute examination and improvement. The first two Midas flights took place on 26 February and 24 May 1960. The first failed during second-stage separation; the second achieved successful orbit and met most of the test objectives. The planned first flight of Samos slipped from April to September 1960.[67]

In November 1959, with the shift of project responsibility for Samos and Midas, OSD reiterated its previous request that the Air Force submit operational plans to the Secretary of Defense for review and approval. These plans would outline the organization structure, operating procedures, and relationships between the operating commands and users of the data derived from the two space systems. The Air Staff prepared the two plans, based in part on preliminary drafts drawn up by the field commands, and AFBMC approved them on 10 February 1960. Two weeks later, Charyk sent them on to OSD.[68]

These plans differed markedly from the usual operational document, being essentially more concept than plan. Briefly, they called for Strategic Air Command to command and operate the Samos system and provide support to the Air Defense Command in the operation of Midas. Continental Air Defense Command would command Midas, with operational control vested in North American Air Defense Command (NORAD).[69]

On 14 March 1960, OSD forwarded both plans to JCS and asked for an early reply, since its comments might affect pending decisions on Samos and Midas development. The Joint Staff readied a draft position by 25 April recommending acceptance of the plans with one major change: Midas should be excepted from NORAD operational control. Army and Navy planners recorded several objections. They favored a joint organization specifically for military space operations,* and they feared that processing and dissemination of Samos-derived data by SAC might lead to inequitable treatment of other users. The USAF planners objected to the exclusion of NORAD from operational control of Midas, claiming that such a step would constitute a breach of American-Canadian agreements. Also it would be a unique arrangement since all other air defense and warning systems operated under NORAD control.[70]

Attempts to settle differences stalled for a few months while JCS addressed the more basic issue of establishing a joint military space operations command. After Secretary Gates reaffirmed on 16 June 1960 the use of existing specified and unified commands for the conduct of military space operations, the Joint Staff reopened the question of Samos and Midas operational assignments and relationships. The Joint Staff generally reiterated its position of 25 April, while the service planners did the same with their earlier objections.[71] At the end of June 1960, it appeared that the issue could be settled only after the Administration and OSD had resolved numerous other facets of the Samos and Midas projects.

---

*See above, pp 18-21.

Notus (Advent)[*]

ARPA established Notus early in 1959 as a two-part communication satellite program, consisting of the interim Courier delayed repeater and the
Decree 24-hour, equatorial-orbiting synchronous systems. Since neither
would meet its requirements for reliable, long-range communications, especially for SAC forces operating in polar regions, the Air Force pressed
for additions to the program. ARPA acceded after several months of discussion and, in May 1959, enlarged Notus by adding Steer and Tackle.
Steer was planned as an interim system using a polar-orbiting, one-channel
communication satellite, while Tackle would be an improved multichannel
system. ARPA designated ARDC (AFBMD) to supervise the development program
under ARPA management and technical direction; the Army's Signal Corps and
the Wright Air Development Center (WADC) would develop payloads and certain other communication equipment as subcontractors to AFBMD.[72]

Dr. York and his staff almost immediately questioned Notus requirements, technical objectives, funding, scope, schedules, and preoccupation
with operational aspects. Beginning in August 1959 and continuing through
April 1960, DDR&E examined all phases of the program. The likely course
of action became discernible on 11 February 1960, when ARPA issued orders
to phase out Courier after completing two scheduled flights, delete Steer
and Tackle immediately, and continue only Decree, redesignated Advent.
The crowning blow to USAF aspirations was the contemplated shift of project supervision from ARPA to the Army instead of the Air Force.[73]

ARPA withdrew the order almost immediately, then on the last day of

---

[*]The active communication satellite program began the year under the
Notus designation; however, following announcement in February 1960 of a
contemplated technical realignment, the revised program was called Advent.

February reissued all but the supervision portion as an interim directive, pending receipt of JCS comments on requirements. JCS reluctantly assented to the reorientation in view of York's gloomy financial report. JCS also affirmed the need for improved communications in the polar region and recommended development of some sort of space system for that purpose.[74]

On 11 April 1960, ARPA released a new order, removing the "interim" tag from the February directive. Advent was now designed basically to demonstrate the technical feasibility of launching a satellite to an orbital altitude of about 20,000 miles, thereby giving it the effect of being stationary, and using it for surface-to-surface communications. For the time being, ARPA eliminated the secondary objective of providing a surface-to-air capability.[75]

Project supervision temporarily remained as before, under ARDC and AFBMD. Subsequently, after AFBMD submitted a development plan that ARPA deemed not particularly responsive to its directive, it appeared only a matter of time until Army would gain project supervision and ultimately take over complete management responsibility from ARPA. On 30 June 1960, Charyk reported that he expected York's decision on this point within several weeks.[76]

Deletion of Steer and Tackle had been based on technical, financial, and scheduling difficulties. Later appeals for reinstatement were unavailing. On 30 June, Charyk explained to General White that if the Air Force regarded the polar communication requirement as sufficiently important—compared with the B-70, Minuteman, Samos, and Midas—to program funds for it, he felt reasonably confident of obtaining reinstatement of the project.[77]

Redirection of the communication satellite program also constituted a major setback to USAF operational objectives. By July 1959, AFBMD had readied a development plan for what it termed the Flag national survival communication satellite, the operational follow-on of the ARPA development program. SAC immediately supported the plan and, to make it more readily acceptable, suggested that the system offered an excellent opportunity to exploit space for peaceful purposes. This could be done by means of a joint effort with private industry, since the latter could use the system for profitable international commercial purposes. If and when an emergency arose, the Air Force could employ the system for military operations.[78]

In October the Air Staff prepared a request for the expenditure of funds to plan and begin construction of operational launching and other essential ground facilities for both the polar (Steer and Tackle) and equatorial (Decree) communication satellites then under development. This type of work required several years of lead time, and the Air Force wanted to be operationally ready as soon as it had demonstrated system feasibility. OSAF sent this request for concurrency to OSD on 8 December 1959, where, not unexpectedly in view of uncertainties in the technical program, it still remained unanswered in July 1960.[79]

Aerospace Defense Systems

The appearance of military ballistic missiles followed by the opening of the space age magnified many old and introduced many new defense problems. Responsibility for research and development of active ballistic defense systems was divided into two parts: the Army-supervised Nike Zeus antimissile missile and ARPA's Project Defender. The latter included a wide variety of studies and applied research on techniques and concepts that might lead to advanced systems for destroying hostile ballistic missiles and satellites. The Air Force, through ARDC, supervised many of these Defender studies and carried out some independent study of its own.

During 1959-60 the Air Force took some first tentative steps toward weapon system development for ballistic missile defense. Deputy Secretary of Defense Gates furnished indirectly the impetus for this action when in October 1959 he indicated that he might soon approve production of Nike Zeus on the premise that no other active defensive system could be available before 1970. The Air Force immediately accelerated planning to show that other and better means could be developed before that date.[80]

The Air Force in its studies of ballistic missile defense had investigated the three possible modes of destroying ballistic missiles: destruction during the boost (powered) phase; destruction during midcourse flight; and destruction during terminal dive. The Army's Nike Zeus would undertake to destroy the missiles in their terminal dive. The Air Force, agreeing with ARPA and DDR&E, was dubious about the effectiveness of this mode. All held similar doubts about destruction in midcourse flight, partly because of the decoy problem. As a result, the Air Force, with ARPA concurrence, concentrated on what appeared to be a most attractive concept--destruction of enemy ballistic missiles in their boost phase. This portion of Defender usually carried an ARPA designation of Bambi (Ballistic Missile Boost Intercept).[81]

After elements of the Air Staff heard briefings on the findings of research to date, DCS/D directed ARDC to prepare an abbreviated development plan to establish the technical feasibility of intercepting ballistic missiles during their boost period. Initial effort was placed on Convair's

Spad (Space Patrol Active Defense) concept: a satellite equipped with in-
frared sensors and containing a large number of tiny interceptors ready to
attack detected missiles. The plan was broadened after Ramo-Woolridge submitted a variation of this idea, called Randon Barrage System (RBS).[82]

Early in January 1960, AFBMD completed the development plan that, if
approved, would supplement current ARPA-sponsored studies and provide a
broader and more timely approach to solution of the defense problem. The
plan called for general design studies, followed by detailed design studies.
Finally, presumably in December 1961, there would be a complete descrip-
tion of a satellite system from which development could proceed to prove
feasibility.[83]

Headquarters USAF readily concurred in the proposal and agreed to reprogram the necessary funds. Charyk, however, felt the plan emphasized
system studies at the expense of the more essential basic investigation
of the many subsystems. In mid-February, after changes had been made to
deemphasize system studies, Charyk approved the plan. Moreover, since the
proposal no longer called primarily for a weapon system, Charyk decided
that DDR&E approval was not needed and that basic investigations could
begin immediately within the USAF applied research program.[84]

ARDC expected to complete these detailed studies during 1960. Assum-
ing favorable results, the Air Force hoped to define within the following
year a system and the component development necessary to start proving the
feasibility of the interception concept. The ultimate objective was to
have an operational satellite system available by 1967.[85]

Study of defensive measures against hostile satellites had begun as
early as 1956 under ARDC sponsorship. In 1958, ARPA assumed responsibility

but continued ARDC as project supervisor. With the steadily advancing space technology, it appeared that a threat in the form of Soviet "bombs in orbit" was possible by 1964. A capability to inspect and, if necessary, destroy any hostile satellite therefore loomed as essential in the near future.[86]

In August 1959, AFBMD, in cooperation with WADC, submitted a preliminary development plan for Saint (Satellite Intercept and Inspection System). The plan called for a program to demonstrate feasibility and develop an operational system in one or more different configurations: unmanned and ground-launched, unmanned and air-launched, and manned. Under existing conditions, ARDC considered it impractical to decide on a single version. Nevertheless, since the Air Staff did not think that it could obtain OSD approval for the proposed broad approach because of financial reasons, DCS/D asked for a revised plan to demonstrate the feasibility of a ground-launched, unmanned, coorbital vehicle possessing rendezvous and inspection capabilities. Other portions of the original proposal would continue to be studied.[87]

The Air Staff, after a thorough review of space defense problems, in December 1959 directed that considerably more emphasis be placed on insuring reliability of Saint subsystems and components prior to fabricating a prototype demonstration system. It carefully reviewed the subsequent plan in February 1960 and then passed it to the Secretary of the Air Force. On 5 April the Air Force asked York's approval to undertake the work as a supplement to ARPA-sponsored studies and agreed to reprogram funds for it. ARPA quickly indorsed the joint effort, and York indicated his favorable views after a 13 May briefing.[88]

Formal DDR&E approval came on 16 June 1960 when York approved the start of a program to demonstrate engineering feasibility of a coorbital satellite system. Work on the prototype Saint system was to be restricted to development, but not flight-testing, of critical subsystems. York asked for yet another revision of the plan based on the above guidelines. As a pleasant bonus, York decided that the Air Force should administer and finance Saint on its own rather than jointly with ARPA.[89]

The new Saint plan was ready on 1 July and quickly gained Air Staff approval. In the AFBMC review on 15 July, Charyk directed that all references to a "kill" capability in the system be eliminated, restricting technical effort to inspection functions only. This step was related to the President's "Space for Peace" program. The plan, now retitled Satellite Inspector System (but still called Saint), went to OSD on 21 July 1960, where it gained approval a month later.[90]

Space Detection and Tracking System

The familiar indecision and controversy characterized management of research and development, assignment of responsibility, and allocation of funds for tracking and identifying vehicles in space. ARPA directed the research and development effort, designated Project Shepherd, and each of the military departments had specific roles to play.

The Air Force portion—Spacetrack—included improvement and operation of the interim data-filtering and -cataloging center, formally termed the Interim National Space Surveillance Control Center (INSSCC), and the preparation of a plan to develop an operational system. Navy activity, called Spasur (Space Surveillance), included development and operation of the minitrack radar net or East-West Fence (constructed originally for Vanguard)

43

and some data-processing facilities in Virginia. The Army would develop
doppler radars and create a network of these—Doploc—to augment Spasur.
ARPA's plan for operating this interim system called for feeding Spasur
and Doploc data, plus that from many other military and civilian sensors
throughout the world, into the INSSCC for reduction, cataloging, position
and trajectory predicting, and dissemination.[91]

At best, the system possessed only a limited capability. York so admitted on 9 March 1960 to a congressional committee. He stated that a good
system would cost between $50 million and $100 million, raising a question
of the urgency and importance of the requirement. DDR&E, ARPA, and the military departments spent much time and effort attempting to arrive at the
proper answer. The Air Force repeatedly proposed development of electronic
scan array radar (ESAR), which it deemed essential for any satisfactory
system, but just as often, ARPA or DDR&E refused approval. Presumably
they were waiting until ARDC's design study on an operational version of
the national space surveillance system, due in ARPA by 30 June 1960, had
been evaluated. Nevertheless, ARPA blasted USAF system objectives on 13
April when it decided that space tracking developments had reached the
point where ARPA no longer needed to support them financially. Any development of improved sensors would require Air Force funds.[92]

The trisegmented operation of the interim tracking system caused fric-
tion among the services, particularly between the Air Force and the Navy.
The Air Force-operated INSSCC was to receive data from the Navy's East-
West Fence and the Army's Doploc network beginning 1 August 1959. Doploc
failed to progress as anticipated and ARPA eventually withdrew its support.
The Navy refused to submit the required data promptly. According to USAF

officials, the Navy delayed and failed to cooperate because it wanted to process the data at its Virginia facility and thereby claim a separate integral system of its own. Only after several meetings and the issuance of ARPA directives did the Navy comply.[93] In addition, the Air Force and Navy tried to obtain management responsibility for the space surveillance system. On each occasion of a formal request for assignment, York indicated that the decision would soon be forthcoming or that reassignment at the time was premature and should remain with ARPA.[94] At the end of June 1960, ARPA still retained management responsibility.

Operational responsibility and organization were frequently discussed but without a decision. In May 1959, McElroy had listed the interim tracking system as one of the four projects sufficiently advanced for reassignment from ARPA and asked for JCS recommendations. As noted above,* this became a part of the joint operational space command controversy. In the split reply that went to McElroy on 24 July 1959, the Army and Navy supported the Joint Staff position calling for operational assignment of the interim space detection system to the proposed joint agency, with the Navy supplying administrative and logistic support and carrying out technical improvement. The Air Force recommended that NORAD assume operational control, with each service supporting its own particular portion of the system. As events transpired, McElroy left the matter undecided, and when he issued his directive of 18 September 1959 on the transfer of management responsibility for four projects the tracking system was not among them.[95]

JCS reopened the subject in April 1960, following receipt of a NORAD request for operational assignment; in fact, the third such request since

_____

*See above, pp 18-19.

1958. From an Air Force point of view, this tended to cloud the issue, since NORAD had been quite inconsistent in its requests. In 1958, NORAD talked of the space detection and tracking system (Spadats); in 1959, it asked for the NSSCC; in 1960, it requested Spacetrack. The three differed considerably—NSSCC was only a part of Spacetrack, and the latter was only a part of the overall Spadats. USAF planners feared that the Army and Navy might exploit these inconsistencies and delay operational assignment.[96]

The Joint Staff initially proposed that, in keeping with McElroy's September directive, the Air Force and Navy draw up operational plans for their respective portions—Spacetrack and Spasur—and forward them to OSD for review and approval. The Navy planners generally agreed, but their Army and Air Force counterparts proposed deferment pending settlement of the joint command question[*] and completion of the technical review currently under way by DDR&E as well as the operational design study due from ARDC by 30 June 1960. JCS agreed to await a decision on the joint command question.[97]

Following Gates' denial on 16 June 1960 of a joint operational structure, the Joint Staff a week later resubmitted most of its original position. The staff tended to go along, however, with the Navy's contention that Spacetrack and Spasur were two independent systems. The Air Force objected, claiming that both were really one system, that each service should continue support of its portion, that the Air Force should receive responsibility for integrating the two parts into an efficient system, and that NORAD should have operational control. This stalemate continued

---

[*]This had been reintroduced by the Navy. See above, pp 20-21.

for several months. The Air Force also tried to gain its objectives through secretarial channels. In May 1960 it sent to ARPA and DDR&E a preliminary concept for the development and operation of the space detection and tracking system, and on 29 June it submitted the 18-month authoritative ARDC study.[98]

Dyna-Soar

For many years the Air Force and the aircraft industry had conducted studies on the feasibility of hypersonic (Mach 5 and above) and orbital flight with a manned vehicle employing boost-glide principles. These studies had been carried out under such project names as Robo, Brassbell, Bomi, and Hywards, finally culminating in Project Dyna-Soar (Dynamic Soaring). Although the Air Force originally conceived Dyna-Soar as a space system, in the fall of 1958 it emphasized suborbital performance for the express purpose of keeping management authority within the Air Force and away from ARPA. Although this maneuver proved successful, ARPA took over direction of the small study effort on an orbital Dyna-Soar weapon system.[99]

Since most of 1959–60 was spent in resolving technical and managerial questions, the Air Force accomplished little in the way of developing Dyna-Soar I, the initial test vehicle. In July 1959 the Air Force completed a competitive contractor evaluation based in part on York's latest guidance, forwarded on 13 April 1959. At that time, York had directed certain fundamental changes. The primary objective of Dyna-Soar I was to be the exploration of hypersonic flight at velocities up to 22,000 feet per second with a vehicle that was manned, maneuverable, launched by a booster already in production or under development, and capable of controlled landings. To be included only if they did not affect schedules or cost or reduce reliability were two secondary objectives: achievement of an orbital capability and provision for installing and testing military subsystems.[100]

The first major obstacle to awarding contracts and starting development appeared on 7 July 1959, when Under Secretary of the Air Force Malcolm A. MacIntyre raised several far-reaching questions on Dyna-Soar plans and directed further study of them. MacIntyre thought that selection of the four-chambered Titan C as the Dyna-Soar booster not only failed to comply with York's directive but that the rocket was too expensive to develop. He also felt that concentration on a vehicle configuration resembling that of the eventual weapon system was premature and too costly. Finally, MacIntyre questioned the proposed management procedures, stressing the necessity for closer relationship within the Air Force between the Wright-Patterson and Inglewood complexes and externally among the Air Force, ARPA, NASA, and industry.[101]

The Air Staff and interested field commands spent the next five months finding answers that would satisfy OSAF and DDR&E, while Dyna-Soar development remained at a standstill. At times it seemed that no answers satisfactory to the several decision-making officials could be furnished. And while the deliberations and discussions were under way, it appeared that OSD with Administration concurrence might withdraw financial support from Dyna-Soar and turn over "bits and pieces" to NASA, which was already participating as a technical adviser.[102]

By mid-November 1959, all the parties concerned were satisfied with the revised development plan and the management procedures. On 17 November, Secretary Douglas and General White approved the program. This included a three-step development plan calling successively for the fabrication and testing of a full-size, 5,000-mile-range glider initially to be airdropped from a B-52 and later ground-launched with a Titan A booster; the extension of glider tests to global range and orbital velocity, using a larger booster, perhaps Saturn; and finally, the development of a weapon system, possibly by 1967. Management responsibility would rest with an augmented ARDC-AMC weapon system project office at Wright-Patterson, which would make extensive use of the experience and knowledge of AFBMC and NASA.[103]

Dyna-Soar had finally moved from what General White had termed "dead center," but a surprise was in store for the Air Staff. On 20 November 1959, Charyk authorized negotiation of contracts with Boeing for the Dyna-Soar system and with Martin for the booster subsystem, but he asked that the financial and work statements be cleared with him before any funds were obligated. On 24 November, Charyk clarified his intentions, disclosing that for the present he would allow only a preliminary study called Phase Alpha. This study would consist of a reexamination of Boeing's proposed technical approach in the light of changes and fund limitations imposed since completion of the competitive study and evaluation in June 1959. Charyk wanted to be doubly sure that the critical aerodynamic, structural, and materials problems so important to the success of Dyna-Soar had been carefully considered.[104]

On 11 December the Air Force contracted with Boeing for one year of development on Dyna-Soar I but restricted work initially to the Phase Alpha study. The findings became available late in March 1960, confirming the previously proposed approach and providing additional confidence in future success. Phase Alpha results were reviewed and approved by succeeding levels of Air Force authority and then sent to York on 19 April. Three days later, York approved the start of Dyna-Soar development and released the required fiscal year 1960 funds. He also emphasized that his directive on objectives, provided one year earlier, remained in effect.[105]

If all went well during the next few years, the first unmanned ground-launching of a Dyna-Soar test vehicle would occur some time between October and December 1963, followed by a manned launching a year later. The results would undoubtedly help crystallize thinking about and formulate characteristics for future manned weapon systems operating in the hypersonic and orbital flight regions.[106]

NOTES

1. *Los Angeles Times*, 8 Dec 59; *Baltimore Sun*, 28 Jan 60; *New York Times*, 17 Mar, 22 Mar, 18 Apr, 15 May, 26 Jun, 12 Jul, & 17 Jul 60; *Washington Post-Times Herald*, 5 Apr & 3 Jun 60.

2. Space Technology Laboratories, STL Space Log, Vol 1, No 1 (1 Jul 60); NASA Office of Tech Info and Educ Progs, draft NASA Hist Report 3, Milestones in Space Exploration, Sep 60.

3. *New York Times*, 21 Dec 59.

4. Msg from the President of the United States ⟨to Congress⟩ Relative to Space Science and Exploration, 2 Apr 58; *New York Times*, 3 Apr 58; NSC 5814/1 (S), 18 Aug 58, in D/Plans Records Br (PRB) files: Sat Prog, RL(59)432, Sec 1.

5. Sources all in PRB files: NSC 5814/1 (S), 18 Aug 58; memo (C) for JCS from Adm J.E. Clark, Actg D/ARPA, 10 Jul 59, subj: Ad Hoc Committee to Review "Preliminary U.S. Policy on Outer Space," in Sat Prog, RL (59)432, Sec 3; memo (S) for C/S from Maj Gen H.T. Wheless, D/Plans, 18 Nov 59, subj: Final Draft of U.S. Policy on Outer Space, in Sat Prog, RL(59)432, Sec 5; JCS 2283/70 (S), 20 Nov 59, in Sat Prog, RL (59)432, Sec 5; memo (S) for SOD from JCS, 23 Nov 59, subj: Final Draft of U.S. Policy on Outer Space, in Sat Prog, RL(59)432, Sec 5; memo (S) for NSC from J.S. Lay, Exec Secy, NSC, 29 Jan 60, subj: U.S. Policy on Outer Space, & NSC 5918 (S), 26 Jan 60, both in Space Prog, RL(60)67-1, Sec 1.

6. NSC 5906/1 (TS), 5 Aug 59, in PRB files: National Security, RL(59)331, Sec 1.

7. NSC 5918 (S), 26 Jan 60.

8. OCB, Operations Plan for Outer Space (S), 15 Jun 60, in PRB files: Space Prog, RL(60)67-1, Sec 2.

9. Msg, President of the United States to the Congress, 14 Jan 60.

10. Rpt, Governmental Organization for Space Activities, 18 Jul 59, by Subcmte on Govt Orgn for Space Activities, Senate Cmte on Aeronautics & Space Sciences, 86th Cong, 1st Sess, 18 Jul 59; Press Release, House Cmte on Science & Astronautics, 29 Oct 59, in OSAF files: Sat Prog, 132-59, Vol 2.

11. Hearings, H Cmte on Science & Astronautics, 86th Cong, 2d Sess, Review of the Space Program (hereinafter cited as Hearings, Brooks Cmte on Space Prog); Hearings, H Cmte on Science & Astronautics, 86th Cong, 2d Sess, To Amend the National Aeronautics and Space Act of 1958 (hereinafter cited as Hearings, Brooks Cmte on Space Act); Hearings, H Subcmte on DOD Appropriations, 86th Cong, 2d Sess, DOD Appropriations for 1961 (hereinafter cited as Hearings, Mahon Cmte); Hearings, S Subcmte on Preparedness Investigating and Cmte on Aeronautics and Space Sciences, 86th Cong, 2d Sess, Missiles, Space and Other Major Defense Matters (hereinafter cited as Hearings, Johnson Cmte); Hearings, S Subcmte on DOD Appropriations, 86th Cong, 2d Sess, DOD Appropriations for 1961 (hereinafter cited as Hearings, Chavez Cmte); New York Times, 19 Jan, 23 Jan, 27 Jan, 14 Feb, & 16 Feb 60; Baltimore Sun, 27 Jan, 31 Jan, & 4 Feb 60; Christian Science Monitor, 11 Feb 60.

12. Hearings, Brooks Cmte on Space Act, pp 80-81, 89-93, 109, 125, 132, 150, 220-21, 525; Draft Agreement, DOD and NASA Concerning the Aeronautics and Astronautics Coordinating Board (AACB), 2 May 60, in D/Aerospace Systems Dev Space Sys Br (SSB) files: Genl.

13. Hearings, Brooks Cmte on Space Act, pp 125, 132; H Rpt 1633, 19 May 60, acted on 9 Jun 60; Draft Agrmt on AACB, 2 May 60; Memo for Record by Col W.E. Gernert, Exec, DCS/D, about 15 Jun 60, subj: Establishment of Joint DOD-NASA Aeronautics and Astronautics Coordinating Board, in SSB files: Spacecraft Panel; Missiles & Rockets, 12 Sep 60.

14. Hearings, Brooks Cmte on Space Prog, pp 167-422, passim.

15. ARPA draft rpt (TS), Advanced Research Plan of DOD, Calendar Years 1959-70; memo (TS) for JCS from N. McElroy, SOD, 10 Aug 59, subj: Long-Range Advanced Research Plan of ARPA; JCS 2283/55/1 (S), 19 Aug 59; memo (S) for SOD from JCS, 27 Aug 59, subj: Long-Range Advanced Research Plan of ARPA, all in PRB files, ARPA, RL(59)10, Secs 1-2; ltr (S), H.F. York, DDR&E to C.F. Ducander, Exec Dir, H Cmte on Science and Astronautics, 30 Dec 59, ns, in SSB files: R&D 18, Long-Range Plng; Hearings, Mahon Cmte, Pt 6, p 7.

16. Hearings, Brooks Cmte on Space Prog, passim; Hearings, Brooks Cmte on Space Act, passim; Hearings, Chavez Cmte, Pt 2, passim; Hearings, Johnson Cmte, passim; Hearings, Mahon Cmte, Pts 1 & 6, passim; Draft Agrmt on AACB, 2 May 60.

17. Hearings, Mahon Cmte, passim; Hearings, Chavez Cmte, passim; H Rpt 1561, 5 May 60; D/Bud, Budget Digests 61 & 65, 7 & 17 Jun 60.

18. Draft Statement on Air Force Policy with Regard to Space, 14 Mar 60, in D/Plans Doctrine Br files.

19. AFM 1-2, United States Air Force Basic Doctrine, 1 Dec 59; interview with Col H.G. McNeese, Ch, Doctrine Br, D/Plans, by author, 14 Mar 61.

20. Memo (S) for DCS/D from Maj Gen H.T. Wheless, D/Plans, 5 Feb 59, subj: Air Force Objectives in Space, in SSB files: R&D 2, Policy; sources in PRB files: ltr (S), Lt Gen B.A. Schriever, Comdr, ARDC to DCS/D, 15 May 59, ns, & ltr (S), Lt Gen J.K. Gerhart, DCS/P&P to Comdr, ARDC, 11 Jun 59, ns, both in Sat Prog, RL(59)432, Sec 3; ltr, Maj Gen H.C. Donnelly, Asst DCS/P&P to C/S, 2 Nov 59, subj: Air Force Concept for the Development and Employment of Space Systems, in Sat Prog, RL(59)432, Sec 5; ltr, Brig Gen J.A. Dunning, Dep D/Plans for War Plans to Dep D/Plans, 10 Feb 60, subj: Status Report, Project 179, in Space Prog, RL(60)67-1, Sec 1; C/S Policy Book, Dec 59.

21. Ltr (C), Maj Gen R.M. Montgomery, Asst VC/S to DCS/P&P, 16 Dec 59, subj: Draft Statement on Air Force Policy with Regard to Space, in PRB files: Sat Prog, RL(59)432, Sec 5; ltr (S), Gerhart to Asst VC/S, 18 Jan 60, subj: Statement of Air Force Policy with Regard to Space, in PRB files: Space Prog, RL(60)67-1, Sec 1; Memo for Record (S) by Col H.G. McNeese, Ch, Doctrine Br, D/Plans, 9 Feb 60, subj: General Estes' Comments on USAF Space Policy, in Doctrine Br files; ltr, Dunning to Dep D/Plans, 10 Feb 60; memo (S) for Asst VC/S from J.V. Charyk, U/SAF, 2 Mar 60, ns, in OSAF files: Sat Prog, 37-69, Vol 1; ltr (S), Donnelly to Asst VC/S, 14 Mar 60, subj: Air Force in Space, in PRB files: Space Prog, RL(60)67-1, Sec 1; intvw with McNeese, 14 Mar 61.

22. Draft Statement on Air Force Policy with Regard to Space, 14 Mar 60.

23. Ltr, Gen T.D. White, C/S to Deps, Dirs, et al., 15 Dec 59, subj: Relations with NASA, & ltr (C), White to DCS/D & DCS/P, 14 Apr 60, ns, both in SSB files: REL 6, NASA.

24. Washington D.C. Evening Star, 13 Dec 59; Hist (S), Bal Msl and Space Sys Div, 1 Jul-31 Dec 60.

25. Hist (S), Bal Msl and Space Sys Div, 1 Jan-30 Jun 60.

26. Ltr (C), White to DCS/D and DCS/P, 14 Apr 60.

27. Ltr, Dunning to Dep D/Plans, 10 Feb 60; Draft Statement on Air Force Policy with Regard to Space, 14 Mar 60.

28. Memo (C) for JCS from McElroy, 18 Sep 59, subj: Coordination of Satellite and Space Vehicle Operations, & memo (S) for SAF from T.S. Gates, Dep SOD, 17 Nov 59, subj: Transfer of the SAMOS Development Program to the Department of the Air Force (similar memos for Midas and Discoverer), both in OSAF files: Sat Prog, 132-59, Vol II; DOD Dir 5129.33, 30 Dec 59; Hearings, NASA Authorization Subcmte, S Cmte on Aeronautics and Space Sciences (Stennis Cmte), 86th Cong, 2d Sess, NASA Authorization for 1961, p 29; memo (C) for SAF from J.H. Douglas, Actg SOD, 6 Feb 60, subj: Transfer of Space Oriented ARPA Projects to the Air Force, in OSAF files: ARPA, 36-60, Vol I.

29. ASSS (C), Brig Gen H.A. Boushey, D/Adv Tech to C/S, 29 Sep 59, subj: Proposed Assignment of ABMA to Department of Air Force, & memo (C) for SAF from Gen C.E. LeMay, VC/S, 29 Sep 59, subj: Proposed Assignment of ABMA to Department of Air Force, both in SSB files: MGT 6, Interservice Relations; memo (C) for D/Plans from Col G.B. Munroe, Tech Exec, D/Adv Tech, 7 Oct 59, subj: ABMA Transfer, in SSB files: Munroe Reading File; memo for JCS from Gates, Actg SOD, 8 Oct 59, subj: Responsibility and Organization for Certain Space Activities, in PRB files: Sat Prog, RL(59)432, Sec 5.

30. Memo for JCS from Gates, 8 Oct 59; memo (S) for SOD from JCS, 13 Oct 59, subj: Responsibility and Organization for Certain Space Activities, in PRB files: Sat Prog, RL(59)432, Sec 5.

31. Memo for President from D/NASA & Actg SOD, 21 Oct 59, ns, in OSAF files: Sat Prog, 132-59, Vol II; JCS 2283/65, 22 Oct 59, & memo (C) for SOD from JCS, 22 Oct 59, subj: Draft DOD--NASA memo for the President, both in PRB files: Sat Prog, RL(59)432, Sec 5; Hist (S), D/Plans, 1 Jul-31 Dec 59.

32. Memo for President from D/NASA & Actg SOD, 21 Oct 59; ltr, Col R.R. Rowland, Dep D/Policy to D/Plans, 13 Nov 59, subj: OSD/NASA (Gates-Glennan) Agreement on Super-boosters and Cost Effectiveness Studies, in PRB files: Sat Prog, RL(59)432, Sec 5.

33. Ltr, LeMay to All AF Pers, 21 Sep 59, subj: Air Force Responsibilities to Other Elements of the DOD for Space Systems, in OSAF files: Sat Prog, 132-59, Vol II; Hearings, Brooks Cmte on Space Prog, pp 474, 484, 494; Hearings, Johnson Cmte, p 59; Memo for Record (S) by McNeese, 9 Feb 60; memo for R.S. Morse, D/R&D, USA, & J.H. Wakelin, Asst SN (R&D) from Charyk, 2 Mar 60, subj: Establishment of a Working Group of Army, Navy, and Air Force Representatives to Confer on Research and Development Space Matters, in SSB files: MGT 1-1, Policy & Directives; Draft Statement on Air Force Policy with Regard to Space, 14 Mar 60.

34. C/S Policy Book, Dec 59; Hearings, Brooks Cmte on Space Prog, p 479.

35. Draft Statement on Air Force Policy with Regard to Space, 14 Mar 60.

36. Memo (C) for JCS from Adm A.A. Burke, CNO, 22 Apr 59, subj: Coordination of Satellite and Space Vehicle Operations; memo (C) for JCS from Gen M.D. Taylor, C/S, USA, 4 May 59, subj: Coordination of Satellite and Space Vehicle Operations; memo (C) for JCS from Gen White, 12 May 59, subj: Coordination of Satellite and Space Vehicle Operations, all in PRB files: Sat Prog, RL(59)432, Sec 2.

37. Memos (S) for JCS from McElroy, 29 May 59, subjs: Assignment of Operational Responsibility for an Interim Satellite Early Warning System; Assignment of Operational Responsibility for an Interim Satellite Navigation System; Assignment of Operational Responsibility for Phase I of a Satellite Reconnaissance System; Assignment of Operational Responsibility for an Interim Satellite Detection System, all in PRB files: Sat Prog, RL(59)432, Sec 2.

38. JCS 2283/32/1 (S), 1 Jun 59; JCS 2283/32/2 (S), 9 Jun 59; memo for
SOD from JCS, 25 Jun 59, subj: Coordination of Satellite and Space
Vehicle Operations; JCS 2283/35/1 (S), 3 Jul 59; JCS 2283/45 (S),
9 Jul 59; memo (S) for JCS from Gen White, 9 Jul 59, subj: Coordination of Satellite and Space Vehicle Operations; JCS 2282/52 (S),
24 Jul 59; memo (S) for SOD from JCS, 24 Jul 59, subj: Coordination
of Satellite and Space Vehicle Operations, all in PRB files: Sat
Prog, RL(59)432, Secs 2-4.

39. Memo for Record (S) by Maj Gen G.W. Martin, Dep D/Plans, 13 Aug 59,
ns, & memo (C) for JCS from McElroy, 28 Aug 59, subj: Coordination
of Satellite and Space Vehicle Operations, both in PRB files: Sat
Prog, RL(59)432, Sec 4; Hist (S), D/Plans, 1 Jul-31 Dec 59.

40. Memo (C) for JCS from McElroy, 18 Sep 59.

41. Memo (C) for JCS from Adm Burke, 4 May 60, subj: Joint Organization
for Command and Control of Military Space Operations, in PRB files:
Space Prog, RL(60)67-1, Sec 1.

42. Memo (C) for C/S from Maj Gen G.W. Martin, Actg D/Plans, 13 May 60,
subj: Joint Organization for Command and Control of Military Space
Operations; memo (C) for C/S from Wheless, D/Plans, 24 May 60, subj:
Joint Organization for Command and Control of Military Space Opera-
tions; JCS 2283/92 (C), 31 May 60; memo for JCS from Gen N.F.
Twining, Chmn, JCS, 28 Jun 60, subj: Joint Organization for Command
and Control of Military Space Operations, all in PRB files: Space
Prog, RL(60)67-1, Sec 1.

43. Memo for JCS & Svc Secys from Gates, 16 Jan 60, subj: Coordination
of Satellite and Space Vehicle Operations, in OSAF files: Sat Prog,
37-60, Vol II.

44. Draft AFOS 2/2 (C), Initial USAF Concept for Space Control, Mar 58;
ASSS, Donnelly to C/S, 26 Mar 58, subj: Objectives Series--Initial
USAF Concept for Space Control; informal memo for SAF from C/S, 31
Mar 58, ns; memo (C) for C/S from J.H. Douglas, SAF, 1 May 58, ns,
all in LR Objs Gp files: AFOS 2/2.

45. Draft AFOS 2/2 (C), Initial USAF Concepts for Operations in Space,
11 Sep 59, in LR Objs Gp files: AFOS 2/2.

46. Ltr, Col J.L. Frisbee, LR Objs Gp to Dep D/Plans for Policy & Dep
D/Plans for War Plans, 14 Sep 59, subj: Initial USAF Concepts for
Operations in Space; ltr (C), Col S.G. Fisher, Asst Dep D/Plans for
Policy to LR Objs Gp, 23 Sep 59, subj: Initial USAF Concepts for
Operations in Space; ltr, Col R.H. Ellis, Asst Dep D/Plans for War
Plans to LR Objs Gp, 29 Sep 59, subj: Initial USAF Concepts for
Operations in Space, all in LR Objs Gp files: AFOS 2/2.

47. Draft AFOS 2/2 (C), 19 Oct & 27 Oct 59, Initial Views on USAF Requirements for Operations in Space; Draft AFOS 2/2 (C), 25 Nov 59 & 20 Jan 60, Initial Concept of USAF Space Activities; ltr, Col Frisbee to D/Dev Plng, 19 Feb 60, subj: AFOS Papers on Space, all in LR Objs Gp files: AFOS 2/2.

48. D/Opnl Rqmts, Required Operational Capability (ROC) for Aerospace, 1970-80 (S), Apr 60, & Presn, ROC for Aerospace, 1970-80, by Col I.J. Klette, Ch, Strat Air Div, D/Opnl Rqmts to DCS/D and D/Sys Mgmt, Jun 60, both in SSB files: R&D.

49. Ltr (S), Col G.B. Munroe, Ch, Policy Div, Asst/Adv Tech to D/Opnl Rqmts, 12 May 60, subj: Draft USAF ROC for 1970-80, in SSB files: R&E; Klette presn, Jun 60.

50. D/Dev Plng, Development Planning Note 59-9 (S), The Air Force in Space, Oct 59, & Revised Development Planning Note 59-9 (S), Mar 60, both in SSB files: R&D.

51. AFCHO study (S), The Threshold of Space, 1945-1959, by Lee Bowen, Sep 60, pp 22-27; memo for D/ARPA from Charyk, 5 Aug 59, subj: Administrative Procedures for ARPA-Sponsored Space Programs, in SSB files: Space Track, Jul-Aug 60; memo for Asst SAF (R&D) from R.W. Johnson, D/ARPA, 19 Aug 59, subj: Administrative Procedures for ARPA-Sponsored Space Programs, & memo for D/ARPA from Charyk, 2 Sep 59, same subj, both in OSAF files: ARPA, 35-59, Vol III.

52. Speech (S) by Lt Gen R.C. Wilson, DCS/D to Air Force Officers Assigned Joint Activities, 17 Mar 60, in SSB files: REL 1, Speeches.

53. Memo (C) for C/S from Douglas, 13 Oct 59, ns, in OSAF files: Sat Prog, 132-59, Vol II.

54. Memo for Record by Col J.L. Martin, Dep D/Adv Tech, 1 Dec 59, subj: Revised Development Plans for SAMOS, MIDAS and DISCOVERER, & ltr, Lt Col F.H. Lucterhand, Asst Exec, DCS/D to D/Adv Tech, 7 Dec 59, same subj, both in AFBMC Sect files: 42d AFBMC Mtg; ltr, Maj Gen R.M. Montgomery, Asst VC/S to Deps et al., 4 Dec 59, subj: Management of Space Matters and Ballistic Missile Matters; ASSS, Wilson to C/S, 7 Dec 59, subj: Staff Procedures for Space Programs; ltr, LeMay to SAF, 23 Dec 59, subj: Staff Procedures for Space Program, all in SSB files: MGT 1-1, Policy Directives.

55. Ltr, Montgomery to Deps et al., 1 Jul 60, subj: Test Procedures for Management of the Ballistic Missile Program, in PRB files: Missile Prog, RL(60)49, Sec 1.

56. Hearings, Brooks Cmte on Space Prog, pp 426-29.

57. Ibid., 476-77; Presn (S) by Brig Gen H.A. Boushey, D/Adv Tech to Special Cmte on the Adequacy of Range Facilities (Cisler Cmte), 18 Sep 59, in SSB files: NSC.

58. Hearings, Brooks Cmte on Space Prog, pp 479-83.

59. Msg (C) DEF 961412, ARPA to ARDC, 23 Jun 59, in OSAF files: Sat Prog, 132-59, Vol I.

60. Memo (S) for D/ARPA from H.F. York, DDR&E, 11 Aug 59, subj: MIDAS Program, in SSB files: Midas, Aug-Sep 59; Amend 7, ARPA Order 38-60 (S), 26 Aug 59, in PRB files: ARPA, RL(59)10; ltr (S), N.E. Golovin, D/Tech Ops, ARPA to Lockheed, 12 Oct 59, ns, in SSB files: Midas, Oct-Dec 59; memo for Actg D/ARPA from E.E. Harriman, Off of D/Tech Ops, ARPA, about 30 Oct 59, subj: Summary of SAMOS Program Situation, & Memo for Record (S) by Maj H.C. Howard, Off of D/Adv Tech, 6 Nov 59, subj: SAMOS, MIDAS and DISCOVERER Programs, both in OSAF-SS files: Samos, Sep-Dec 59; ltr (S), Maj H.C. Howard, Chmn, Midas Working Gp to Midas Working Gp, 11 Dec 59, subj: Point 6 Program, in SSB files: Midas, Oct-Dec 59; Min (S) of 42d AFBMC Mtg, 14 Dec 59, in AFBMC Sect files: 42d AFBMC Mtg; msg (S) 98212, Hq USAF to ARDC, 21 Dec 59, & ltr (S), Wilson to C/S, 19 Jan 60, subj: Status of SAMOS, MIDAS and DISCOVERER, both in SSB files: Midas, Oct 59-Feb 60; ltr (C), Wilson to Comdr, ARDC, 1 Mar 60, subj: Reliability Management, in OSAF-SS files: Samos, Mar-Apr 60; ltr (C), Wilson to Comdr, ARDC, 1 Mar 60, subj: Reliability Analysis--Correlation of Actual Experience with Theoretical Studies, in SSB files: Samos, Mar-Apr 60; Hearings, Chavez Cmte, Pt 2, p 1207; ltr (S), Wilson to Comdr, ARDC, 1 Jun 60, Exploitation of Initial SAMOS Data, in OSAF-SS files: Samos, May-Jun 60.

61. Ltr (S), Schriever to C/S, 1 Aug 59, ns, in OSAF-SS files: Samos Funds, 1959; ltr (S) LeMay to Comdr, ARDC, 9 Sep 59, ns; ltr (S), Schriever to C/S, 15 Sep 59, ns /Samos/; ltr (S), Schriever to C/S, 15 Sep 59, ns /Midas/; ltr (S), LeMay to Comdr, ARDC, 13 Oct 59, ns, all in OSAF-SS files: Samos, Sep-Oct 59; msg (S) VC-5540, SAC to Hq USAF, 16 Dec 59, in SSB files: Midas, Oct-Dec 59; ltr (S), Maj Gen J.H. Walsh, ACS/I to DCS/D, 21 Dec 59, subj: SAMOS, in DCS/D files: R&D-8, Missiles; msgs (S): 98219, Hq USAF to SAC, 21 Dec 59; ADLPD-D-1, ADC to Hq USAF, 4 Jan 60; 61416, Hq USAF to ADC, 7 Jan 60; VC-0206, SAC to Hq USAF, 9 Jan 60, all in SSB files: Midas, Dec 59-Feb 60; msg (S) C-0706, SAC to Hq USAF, 25 Jan 60, in OSAF-SS files: Samos, Jan-Feb 60; msg (S) VC-0847, SAC to Hq USAF, 28 Jan 60, & msg (S) /AFIN 58998/, ADC to Hq USAF, 30 Jan 60, both in SSB files: Midas, Jan-Feb 60; msg (S) 67350, Hq USAF to SAC, 1 Feb 60, in OSAF-SS files: Samos, Jan-Feb 60; msg (C) 69530, Hq USAF to ADC, 10 Feb 60, in SSB files: Midas, Jan-Feb 60; ltrs (S), Wilson to Schriever, 1 Mar 60.

62. Samos Dev Plans, dtd 15 Jul 59, 1 Dec 59, 15 Jan 60, & 12 Jul 60; Midas Dev Plans, dtd 15 Jul 59, 1 Dec 59, & 15 Jan 60; Discoverer Dev Plans, dtd 15 Jul 59, 1 Dec 59, & 15 Jan 60, plus several other unpublished drafts.

63. AFBMD, Samos Dev/Ops Plan (S), 15 Jan 60, in OSAF-SS files: Samos, Jan–Feb 60; AFBMD, Midas Dev/Ops Plan (S), 15 Jan 60, in SSB files: Midas, Jan–Feb 60; memo (S) for SOD from Charyk, 18 Feb 60, subj: Transfer of the SAMOS, MIDAS and DISCOVERER Programs to the Department of the Air Force, in OSAF files: Sat Prog, 132-59, Vol II; memo (S) for SAF from York, DDR&E, 20 Apr 60, subj: SAMOS, MIDAS and DISCOVERER Research and Development Programs and Development/Operational Plans for SAMOS and MIDAS Programs, in OSAF files: Sat Prog, 37-60, Vol 1.

64. Ltr (S), Wilson to D/Adv Tech, 9 May 60, subj: SAMOS, in OSAF-SS files: Samos, May–Jun 60; ltr (TS), Wilson to Asst SAF (R&D), 9 May 60, subj: Project SAMOS, in OSAF files: Sat Prog, 37-60; ltr (S), Maj Gen V.R. Haugen, Asst DCS/D to Comdr, ARDC, 16 May 60, subj: Samos Development Program; ltr (S), Wilson to ARDC, 1 Jun 60, as cited in n 60; ltr (S), White to Comdrs, ARDC, SAC, & ADC, 29 Jun 60, subj: Supplemental Hq USAF Guidance to ARDC, SAC and ADC Concerning SAMOS; ltr (S), Col R.B. Allison, Exec, C/S to VC/S, 2 Jul 60, subj: SAMOS, all in OSAF-SS files: Samos, May–Aug 60.

65. New York Times, 13 Jun 60; Aviation Daily, 14 Jun 60; ASSS, Maj Gen R.J. Friedman, D/Bud to C/S, 12 Jul 60, subj: Congressional Add-Ons, & draft memo for DDR&E prep by Asst/Adv Tech, 19 Jul 60, subj: Establishing Our Staff Recommendations for the "Add-Ons," both in OSAF-SS files: Samos, Jul–Aug 60.

66. USAF Current Status Rpt (S), Jul 59 thru Jun 60; Hearings, Chavez Cmte, Pt 2, p 1067.

67. Hist (S), D/Adv Tech, Jul–Dec 59 & Jan–Jun 60.

68. Memo (S) for SAF from Gates, 17 Nov 59, as cited in n 28; Min (S) of 45th AFBMC Mtg, 10 Feb 60; ASSS (S), Brig Gen J.K. Hester, Dep D/Ops to SAF, 25 Feb 60, subj: Preliminary Operations Plans for SAMOS and MIDAS; memos (S) for SOD from Charyk, 25 Feb 60, subjs: Preliminary Operations Plan for the Missile Defense Alarm System, MIDAS, & Preliminary Operations Plan for the Satellite Reconnaissance System, SAMOS, all in OSAF-SS files: Samos, Jan–Feb 60.

69. See n 68.

70. Memo for JCS from Douglas, 14 Mar 60, subj: Preliminary Operational Plans for SAMOS and MIDAS, in OSAF files: Sat Prog, 37-60, Vol I; JCS 2283/82/1 (S), 25 Apr 60, & memo (S) for Dep D/Plans from Col N.S. Orwat, Asst Dep D/Plans for War Plans, 4 May 60, subj: Preliminary Operations Plan for SAMOS and MIDAS, both in PRB files: SAMOS Prog, RL(60)67-3, Sec 1.

71. Memos (S) for Dep D/Plans from Col Orwat, 22 Jun & 1 Jul 60, subj of both: Preliminary Operations Plan for SAMOS and MIDAS, both in PRB files: SAMOS Prog, RL(60)67-3, Sec 1.

72. ARPA Order 54-59 (S), 20 Jan 59; ARPA Order 55-59 (S), 20 Jan 59; GOR 178 (S), 2 Feb 59; Memo for Record (S) by Maj H.C. Howard, Off of D/Adv Tech, 9 Feb 59, subj: Communication Satellite; memo for Deps et al. from LeMay, 6 Mar 59, subj: Air Force Position on Communications Satellite Programs; Amend 1, ARPA Order 54-59 (S), 22 May 59, all in SSB files: Comm Sat, Jan-Sep 59.

73. Memo (S) for DCS/D from Boushey, 17 Aug 59, subj: SAC Polar Communications Satellite, in SSB files: Boushey Correspondence; msg (C) 68749, Hq USAF to SAC, ARDC, & AFBMD, 28 Aug 59; memo (C) for ARDC from N.E. Golovin, D/Tech Ops, ARPA, 21 Oct 59, subj: Technical Guidance, Communication Satellite Program; msg (S) 87347, Hq USAF to ADC, 29 Oct 59; msg (C) WDZSC-11-2-E, AFBMD to Hq USAF, 5 Nov 59; msg (S) 89129, Hq USAF to AFBMD, 6 Nov 59; msg (C) DEF 96889, ARPA to ARDC, 20 Nov 59; ltr (C), Col H.L. Evans, Asst Dep Comdr, AFBMD to C/S, 27 Nov 59, subj: Objectives of Project Steer; msg (S) 96368, Hq USAF to SAC, 11 Dec 59; Amend 4, ARPA Order 54-60 (C), 11 Feb 60, & Amend 2, ARPA Order 55-60 (C), 11 Feb 60, both of which were withdrawn immediately, all in SSB files: Comm Sat, Aug 59, Oct-Dec 59, & Jan-Mar 60; memo (C) for DDR&E from Charyk, 21 Jan 60, subj: Satellite Communications Systems-Project STEER, in OSAF files: Sat Prog, 132-59, Vol II.

74. Amend 4, ARPA Order 54-60 (C), 29 Feb 60, in SSB files: Comm Sat, Jan-Mar 60; memo (S) for JCS from Douglas, 14 Mar 60, subj: Reorientation of Communications Satellite Research and Development Program, & memo (S) for SOD from JCS, 31 Mar 60, subj: Reorientation of Communications Satellite Research and Development Program, both in PRB files: Space Prog, RL(60)67-1, Sec 1.

75. Amend 5, ARPA Order 54-60 (C), 11 Apr 60, in SSB files: Comm Sat, Apr-Jun 60.

76. Draft Advent Dev Plan (S), 3 Mar 60; Advent Dev Plan (S), 1 Apr 60; Memo for Record (S) by Col G.B. Munroe, Ch, Policy Div, Asst/Adv Tech, 12 Apr 60, subj: Communications Satellite; draft ltr (C), not sent, Col J.L. Martin, Dep Asst/Adv Tech to DCS/D, 26 Apr 60, subj: Management of the Communications Satellite Program; Memo for Record (C) by Col Munroe, 4 May 60, subj: ARPA Briefing on AFBMD Communications Satellite Development Plan--4 May 1960; memo (C) for C/S from Charyk, 30 Jun 60, subj: Satellite Communications Systems, all in SSB files: Comm Sat, Apr-Jun 60.

77. ASSS (C), Maj Gen H.W. Grant, D/C-E to SAF, 9 Jun 60, subj: Satellite Communications Systems, in OSAF files: Sat Prog, 37-60, Vol II; memo for C/S from Charyk, 30 Jun 60.

78. D/C-E, Preliminary Opnl Concept for Polar Orbit Communications Satellite System (C), nd but about 16 Jul 69; AFBMD, National Survival Communications Satellite (FLAG) Dev Plan (S), 27 Jul 59; ltr (S), Lt Gen F.H. Griswold, V/CinC, SAC to C/S, 18 Aug 59, subj: National Survival Communications Satellite; ltr (S), Schriever to C/S, 15 Sep 59, subj: National Communications Satellite Program, all in SSB files: Comm Sat, Jan-Sep 59.

79. ASSS (C), Grant to Asst SAF (Mat), 29 Oct 59, subj: Satellite Communications Systems Program Objectives for the Air Force, & ltr (C), Grant to Deps et al., 30 Oct 59, subj: Approved Preliminary Operational Concepts for Satellite Communications Systems; memo (C) for Asst SOD (S&L) from P.B. Taylor, Asst SAF (Mat), 8 Dec 59, subj: Satellite Communications Systems Program Objectives for the Air Force, all in SSB files: Comm Sat, Oct-Dec 59.

80. Ltr (S), Col W.H. Earle, Ch, Air Def Div, D/Dev Plng to Comdr, ARDC, 23 Oct 59, subj: Active Ballistic Missile Defense; memo (S) for Asst D/Adv Tech from Maj H.C. Howard, Off of D/Adv Tech, 29 Oct 59, subj: AICBM Briefing to Air Defense Panel; Rpt (S) of 76th Wpns Bd Mtg, 2 Nov 59, all in SSB files: SCWS.

81. Sources in SSB files: Rpt (S) of 76th Wpns Bd Mtg, 2 Nov 59; SDR 14 (S), 20 Nov 59, in MGT 5-2, SDR's; Presn (S), Space Defense, by Col Earle to Wpns Bd, 16 Mar 60, in Saint, 1960; ARPA Rpt (S), Review of Project Defender, 25-29 Jul 60, in Space Track, Jul-Aug 60.

82. Memo (S) for Asst D/Adv Tech from Howard, 29 Oct 59, as cited in n 80; Rpt (S) of 76th Wpns Bd Mtg, 2 Nov 59; SDR 14 (S), 20 Nov 59; ltr (C), Lt Col D.P. Andre, Exec, D/Dev Plng to DCS/P&P, 9 Nov 59, subj: ICBM Defense Policy, & ltr (C), Maj Gen V.R. Haugen, D/Dev Plng to Comdr, ARDC, 23 Dec 59, subj: Ballistic Missile Defense, both in SSB files: SCWS.

83. AFBMD, Orbital Interceptor Dev Plan (S), 7 Jan 60, in SSB files: Spad; ltr (C), Maj Gen J. Ferguson, V/Comdr, ARDC to C/S, 12 Jan 60, subj: Abbreviated Development Plan—SDR #14; Consolidated Minutes of AAICBMD /Active Anti-Intercontinental Ballistic Missile Defense/ Working Group (S), 3 Dec 59-12 Jan 60, both in SSB files: SCWS.

84. Memo (S) for DCS/D from Charyk, 8 Feb 60, ns, in D/Dev Plng Air Def Div (ADD) files: SCWS; ltr (S), Haugen to DCS/D, 10 Feb 60, subj: Advanced Ballistic Missile Defense, & ltr (S), Maj Gen L.I. Davis, Asst DCS/D to Asst SAF (R&D), 11 Feb 60, subj: Advanced Ballistic Missile Defense, both in SSB files: SCWS; ltr (S), Earle to Asst SAF (R&D), 16 Feb 60, subj: Advanced Ballistic Missile Defense, in ADD files: SCWS; ltr (S), Col B.R. Lawrence, Dep D/R&D to Comdr, ARDC, 29 Feb 60, subj: Approval for Work under SDR dated 24 November 1959, in SSB files: SCWS.

85. Orbital Interceptor Dev Plan (S), 7 Jan 60; USAF Current Status Rpt (S), Advanced AICBM, Apr 1960, p 98.

86. ARDC SR 143 (S), 28 Aug 56; ARDC SR 187 (S), 1 May 58; ARDC Project 7991 Mgmt Rpt (S), 30 Jun 59, all in SSB files: Saint, 1958-59; ARPA Rpt (S), Review of Project Defender, 25-29 Jul 60.

87. AFBMD, Saint Dev Plan (S), 10 Aug 59; ltr (S), Lt Col C. Arnold, Dep Asst/Programming, AFBMD to C/S, 21 Aug 59, subj: Satellite Intercept and Inspection System; ltr, Col Martin, Dep D/Adv Tech to Comdr, ARDC, 2 Sep 59, subj: Satellite Intercept and Inspection System; ltr (S), Schriever to C/S, 15 Sep 59, subj: Satellite Intercept and Inspection System; msg (S) 79042, Hq USAF to AFBMD, 1 Oct 59, all in SSB files: Saint, 1958-59.

88. Rpt (S) of 76th Wpns Bd Mtg, 2 Nov 59; ltr (S), Haugen to Comdr, ARDC, 7 Dec 59, subj: Satellite Intercept and Inspection System; ltr (S), Earle to Comdr, ARDC, 26 Jan 60, subj: Satellite Intercept and Inspection System; AFBMD, Saint Dev Plan (S), 8 Feb 60; ltr (S), Haugen to DCS/D, 25 Mar 60, subj: SAINT Proposal; AFC 15/23 (S), 25 Mar 60; ltr (S), Wilson, DCS/D to Asst SAF (R&D), 28 Mar 60, subj: Satellite Inspection; memo (S) for ARPA and DDR&E from C.D. Perkins, Asst SAF (R&D), 5 Apr 60, subj: Satellite Inspection, all in SSB files: Saint 1959-60; ltr (S), Wilson to C/S, 18 May 60, subj: R&D Management, in SSB files: MGT-6, R&D Mgmt.

89. Memo (C) for Asst SAF (R&D) from York, 16 Jan 60, subj: Satellite Inspection, in OSAF files: Sat Prog, 37-60, Vol II.

90. AFBMD, Saint Dev Plan (S), 1 Jul 59, in SSB files: Saint, 1960; Min (S) of 54th AFBMC Mtg, 15 Jul 60, & memo for DDR&E from Perkins, 21 Jul 60, subj: Development Plan for SAINT, both in AFBMC Sect files: 54th AFBMC Mtg; memo (C) for Asst SAF (R&D) from J.H. Rubel, Dep DDR&E, 25 Aug 60, subj: Satellite Inspection, in OSAF files: Sat Prog, 37-60, Vol II.

91. ARPA Order 50-59 (C), 9 Dec 58, in SSB files: ARPA Orders; AFCRC, Space Track Sys Dev Plans (S), 17 Jul 59, in SSB files: Space Track, Jul-Aug 59.

92. Hearings, Mahon Cmte, Pt 6, p 123; Memo for Record (C) by D. Duke & D.C. Holmes, ARPA, 23 Oct 59, subj: Air Force Procurement of Phased Array Radar Systems for Project SPACE TRACK, & ltr (S), Lt Col J.W. Lillard, DCS/R&E, ARDC to C/S, 2 Dec 59, subj: 496L System Development Plan, both in SSB files: Space Track, Oct-Dec 59; ASSS (C), Col J.R.V. Dickson, D/Dev Plng to SAF, 15 Mar 60, subj: Request for Allocation of Funds from the Secretary of Defense Emergency Funds, & memo (C) for D/Adv Tech from J.P. Puina, Dep/Research, Asst SAF (R&D), 28 Mar 60, subj: Emergency Fund Request for Space Track, both in SSB files: Space Track Funding; Amend 5, ARPA Order 50-60, 13 Apr 60; msg (C) 89036, Hq USAF to ARDC, 26 Apr 60; msg (C) RDSPE-7/2-5-1, ARDC to Hq USAF, 2 May 60, all in SSB files: Space Track, Mar-Jun 60.

93. ARPA Order 50-59 (C), 19 Dec 58; Amend 4, ARPA Order 7-58 (C), 21 Apr 59; ltr, Brig Gen M.F. McNickle, Dep Comdr, ARDC to C/S, 15 Sep 59, subj: Integration of the A-N Fence with the Air Force Space Track System, & memo (C) for D/NRL and Ch, 496L ESSPO from P.A. Price, Tech Ops Div, ARPA, 8 Oct 59, subj: October 5, 1959 Meeting on Furnishing of NRL Dark Satellite Fence Data to SPACE TRACK, both in SSB files: Space Track, Jun-Oct 59; AFCRC, Space Track Mo Progress Rpt (C), Aug 59, Oct 59, Dec 59, & (S) Feb 60, in SSB files.

94. ASSS (S), Boushey to Asst SAF (R&D), 18 Dec 59, subj: Transfer of NSSCC to the Department of the Air Force, in SSB files: Space Track, Nov-Dec 59; memo (S) for DDR&E from Charyk, 29 Jan 60, subj: Transfer to the Department of the Air Force of Responsibility for Development of the National Space Surveillance Control Center, in OSAF files: Sat

Prog, 132-59, Vol II; memo (C) for DDR&E from J.H. Wakelin, Asst SN
(R&D), 11 Feb 60, subj: Responsibility for Management and Development
of the Space Surveillance System, in SSB files: Space Track, Jan-Feb
60; memo (C) for DDR&E from Charyk, 16 Feb 60, subj: Responsibility
for Management and Development of a Space Surveillance System, & memo
(S) for U/SAF from York, 19 Mar 60, same subj, both in OSAF files:
Sat Prog, 37-60, Vol I; memo (S) for ARPA from H. Davis, Dep/Research,
Asst SAF (R&D), 10 Jun 60, subj: Concept for Organization, Functions,
and Management of a National Space Detection and Tracking System, in
SSB files: Space Track, May-Jun 60.

95. Memo for JCS from McElroy, 29 May 59, subj: Assignment of Operational
Responsibility for an Interim Satellite Detection System; memo for
SOD from JCS, 24 Jul 59; memo for JCS from McElroy, 18 Sep 59, all as
cited in notes 37-38 & 28.

96. Hist (S), D/Plans, Jul-Dec 59 & Jan-Jun 60; ltr (S), Gen L.S. Kuter,
CinC, NORAD to JCS, 20 Apr 60, subj: Assignment of Operational Responsibility for Satellite Detection and Tracking System, in PRB
files: Space Prog, RL(60)67-1, Sec 1.

97. JCS 2283/87/1 (S), 20 May 60; JCS 2283/87/2 (S), 1 Jun 60; memo (S)
for Dep D/Plans from Col N.S. Orwat, Asst Dep D/Plans for War Plans,
8 Jun 60, subj: Assignment of Operational Responsibility for Satellite Detection and Tracking System, all in PRB files: Space Prog,
RL(60)67-1. Sec 1.

98. JCS 2283/3 (S), 22 Jun 60, & memo (S) for Dep D/Plans from Orwat, 29
Jun 60, subj: Assignment of Operational Responsibility for Satellite
Detection and Tracking System, both in PRB files: Space Prog, RL(60)
67-1, Sec 1; Asst/Adv Tech, Concept for Organization, Functions and
Management of a Space Detection and Tracking System (S), May 1960, &
496L ESSPO Rpt (S), Conceptional Design Plan for the National Space
Surveillance System (NSSS) . . . , Jul 60, both in SSB files: Space
Track, May-Jun 60.

99. AFCHO study, The Threshold of Space, pp 35-36; ARPA Order 84-59, 30
Apr 59, in SSB files: Dyna Soar Policy Dirs, Apr-Dec 59.

100. Memo (S) for SAF from York, 13 Apr 59, subj: Dyna-Soar I Program
Guidance, in SSB files: Dyna Soar Policy Dirs, Apr-Dec 59; memo (S)
for C/S from M.A. MacIntyre, Actg SAF, 7 Jul 59, subj: Dyna Soar
Source Selection, in PRB files: Sat Prog, RL(59)432, Sec 3.

101. Memo for C/S from MacIntyre, 7 Jul 59.

102. Msg (S) 53166, Hq USAF to Det 1, ARDC, 10 Jul 59; memo for SAF and
D/ARPA from York, 27 Jul 59, subj: Saturn-Dyna Soar Propulsion; Memo
for Record (S) by Lt Col B.H. Ferer, Off of D/Adv Tech, 29 Sep 59,
subj: DYNA SOAR Division; ltr (C), Davis, Asst DCS/D to C/S, 1 Oct
59, subj: Status of DYNA SOAR Project; ASSS, Boushey to SAF, 23 Oct
59, subj: Dyna Soar Development; ltr (S), Boushey to DCS/D, 29 Oct

59, subj: OSD—NASA Meeting on Configuration of Saturn; memo (S) for C/S from Wilson, 29 Oct 59, Dyna Soar Development Plan; ltr (S), Gen S.E. Anderson, Comdr, AMC & Schriever, Comdr, ARDC to C/S, 29 Oct 59, subj: Dyna Soar Source Selection; ltr (C), Boushey to C/S, 30 Oct 59, subj: Statement of Critical Problems Concerning SATURN, DYNA SOAR, and Air Force Space Responsibilities; Det 1, ARDC, Abbreviated Dyna Soar Dev Plan (S), 6 Nov 59; Memo for Record by Ferer, 10 Nov 59, subj: Actions on Dyna Soar—5 Sep 59 thru 6 Nov 59, all in SSB files: Dyna Soar Policy Dirs, Apr–Dec 59.

103. AFC 1/4a (S), 17 Nov 59, & msg (S) 90938, Hq USAF to AMC and ARDC, 17 Nov 59, both in SSB files: Dyna Soar Policy Dirs, Apr–Dec 59.

104. Memo (S) for DCS/D, DCS/M, & D/Bud from Charyk, 20 Nov 59, subj: Dyna-Soar Prog; Memo for Record (C) by Ferer, 24 Nov 59, subj: Asst SAF R&D's Restriction on DYNA SOAR Determinations and Findings; memo (C) for DCS/D, DCS/M, & Compt from Charyk, 7 Dec 59, subj: Dyna-Soar Program, all in SSB files: Dyna Soar Policy Dirs, Apr–Dec 59.

105. USAF Current Status Rpt (S), 464L-DYNA SOAR, 18 Dec 59, & 620A-DYNA SOAR, Apr 60; Rpt of Wpns Bd Mgt 60-17 (S), 8 Apr 60; AFC 15/24, 13 Apr 60; memo (S) for DDR&E from C.D. Perkins, Asst SAF (R&D), 19 Apr 60, subj: Approval of Funds for Dyna Soar; memo (C) for SAF from York, 22 Apr 60, subj: Dyna-Soar I Program, all in SSB files: Dyna Soar Policy Dirs, 1960.

106. USAF Current Status Rpt (S), Apr 60.

GLOSSARY

AACB	Aeronautics and Astronautics Coordinating Board
ABMA	Army Ballistic Missile Agency
AEC	Atomic Energy Commission
AFBMC	Air Force Ballistic Missile Committee
AFBMD	Air Force Ballistic Missile Division
AFC	Air Force Council
AFCHO	USAF Historical Division Liaison Office
AFCRC	Air Force Cambridge Research Center
AFOS	Air Force Objective Series
AMC	Air Materiel Command
ARDC	Air Research and Development Command
ARPA	Advanced Research Projects Agency
ASSS	Air Staff Summary Sheet
Bal	Ballistic
C-E	Communications-Electronics
CMLC	Civilian-Military Liaison Committee
CNO	Chief of Naval Operations
D/	Director of
DDR&E	Director of Defense Research and Engineering
ESSPO	Electronic Support System Project Office
INSSCC	Interim National Space Surveillance Control Center
Intvw	Interview
JCS	Joint Chiefs of Staff
Msl	Missile
NASA	National Aeronautics and Space Administration
NASC	National Aeronautics and Space Council
NRL	Naval Research Laboratory
ns	no subject
NSC	National Security Council
NSSCC	National Space Surveillance Control Center
Objs	Objectives
OCB	Operations Coordinating Board
Opnl	Operational
Ops	Operations
OSAF-SS	Office, Secretary of the Air Force, Samos files
OSD	Office, Secretary of Defense

PRB	Records Branch, Directorate of Plans
Presn	Presentation
Prog	Program
RBS	Random Barrage System
ROC	Required Operational Capability
Saint	Satellite Intercept and Inspection System; Satellite Inspector System
Sat	Satellite
SN	Secretary of the Navy
SOD	Secretary of Defense
Spad	Space Patrol Active Defense
Spadats	Space Detection and Tracking System
Spasur	Space Surveillance
SSB	Space Systems Branch, Directorate of Aerospace Systems Development
STL	Space Technology Laboratory
Svc	Service
U/SAF	Under Secretary of the Air Force
WADC	Wright Air Development Center

www.ingramcontent.com/pod-product-compliance
Lightning Source LLC
Chambersburg PA
CBHW081237090426
42738CB00016B/3340